Physics I
Mechanics
Exam File

Arthur B. Western, Montana College of Mineral Science and Technology, Editor; H. S. Ahluwalia, The University of New Mexico; Dwight B. Beery, Manchester College; Treasure Brasher, West Texas State University; Rufus E. Bruce, The University of Texas at El Paso; Robert Chasnov, Liberty University; Gayl Cook, University of Colorado at Denver; Roger C. Crawford, Pierce College; J. W. Culvahouse, The University of Kansas; Bruce I. Ewen, Wentworth Institute of Technology; Mason Fisher, Youngstown State University; Leon P. Goldberg, Glassboro State College; Warren W. Hein, South Dakota State University; Dennis C. Henry, Gustavus Adolphus College; Paul N. Houle, East Stroudsburg University; Steven M. Kahn, University of California, Berkeley; Jack R. Leibowitz, Catholic University; Jerome R. Long, Virginia Polytechnic Institute and State University; John L. McClure, Prince George's Community College; William C. McGowan, Clarion University; T. Ted Morishige, Central State University; David L. Mott, New Mexico State University; Bill Mundy, Pacific Union College; Paul J. Nienaber, Eastern Illinois University; Allan B. Packman, University of Hartford; James J. Peters, Hillsdale College; Otto I. Reisman, New Jersey Institute of Technology; Sedgwick Simons, University of Houston Downtown College; John J. Sinai, University of Louisville; Ralph Snyder, The University of Connecticut; Jay D. Strieb, Villanova University; Willem van den Berg, The Pennsylvania State University, DuBois Campus; E. J. Zimmerman, The University of Nebraska - Lincoln

ENGINEERING PRESS, INC.   SAN JOSE, CALIFORNIA 95103-0001

Donald G. Newnan, Ph.D.
EXAM FILE Series Editor

©Copyright 1986, Engineering Press, Inc.
All rights reserved. Reproduction or
translation of any part of this work beyond
that permitted by section 107 or 108 of the
1976 United States Copyright Act without the
permission of the copyright owner is
unlawful.

Printed in the United States of America
5 4 3 2 1

ISBN 0-910554-54-4

Library of Congress Cataloging-In-Publication Data

Main entry under title:

Physics exam file.

    (Exam file series)
    Contents: 1. Mechanics -- 2. Heat, light, and sound -- v. 3 Electricity and magnetism.
    1. Physics--Problems, exercises, etc.--Collected works. 2. Physics--Examinations,
questions, etc--Collected works. I. Western Arthur B. (Arthur Boyd), 1944-   . II.
Ahluwalia, H. S. (Harjit Singh) III. Series.
QC32-P46   1986   530'.076   85-25306
ISBN 0-910554-54-4 (v. 1.)
ISBN 0-910554-55-2 (v. 2)

Engineering Press, Inc.   P.O. Box 1   San Jose, California 95103-0001

Contents

    Foreword   v

1. VECTORS
    Addition and Subtraction   1
    Components and Resultants   3
    The Dot and Cross Products   5

2. VELOCITY AND ACCELERATION
    Average Velocity   13
    Instantaneous Velocity   15
    Average Acceleration   17
    Instantaneous Acceleration   18

3. EQUILIBRIUM
    Newton's First Law   31
    Equilibrium of a Particle   32
    Frictional Forces   38
    The Moment of a Force - Torque   42
    Center of Gravity   44
    Equilibrium of Rigid Bodies   46
    Frictional Forces   62

4. PLANE MOTION
    Newton's Second Law   63
    Constant Acceleration   82
    Projectile Motion   96
    Circular Motion   113
    Gravitation
        Newton's Law of Gravitation   128
        Satellite Motion   132
    Motion in a Vertical Circle   140
    Relative Velocity   142

5. WORK AND ENERGY
Work Done by a Constant Force	149
Work Done by a Variable Force	155
Kinetic Energy	160
Gravitational Potential Energy	165
Conservation of Energy	169
Elastic Potential Energy	183

6. IMPULSE AND MOMENTUM
Impulse and Momentum	193
Conservation of Linear Momentum	201
Elastic Collisions	207
Inelastic Collisions and the Coefficient of Restitution	213
Variable Mass Systems	233

7. ROTATIONAL DYNAMICS
Angular Displacement, Velocity, and Acceleration	235
Rotation With Constant Angular Frequency	239
Moment of Inertia	240
Torque and Angular Acceleration	245
Rotation With Constant Angular Acceleration	249
Rotation with Translation	256
Angular Momentum and Impulse	258
Conservation of Angular Momentum	260
Kinetic Energy, Work and Power	264
Gyroscopic Effects	281

8. ELASTICITY AND HARMONIC MOTION
Stress, Strain, and the Elastic Moduli	283
Hooke's Law and Simple Harmonic Motion	286
Energy Observations with Simple Harmonic Motion	298
Damped Harmonic Motion	305
The Simple Pendulum	306
The Physical Pendulum	311

9. FLUIDS
Pressure and Density in Fluids	315
Archimedes' Principle	318
Pascal's Principle	333
Forces Against a Dam	334
The Equation of Continuity	335
Bernoulli's Equation	337

Foreword

This book is the first of a three volume set designed to accompany an introductory, college-level, calculus-based, physics course. Students enrolled in such courses often encounter two difficulties: (1) learning to use mathematics as a description of the physical world and (2) preparing for tests which stress the *application* of relatively few laws, rather than rote memorization of a multitude of facts. Happily both difficulties have the same solution, namely, doing more problems. But, the student may answer quickly, how can I do more problems when my whole difficulty is that I can't do the problems? The way out of this dilemma is to understand the problem solutions of others and to attempt to apply the same techniques oneself. The purpose of this book is to make available suitable problems and solutions which will aid the student in both learning physics and passing tests.

Professors from around the country have agreed to open their exam files and allow their problems and solutions to be published. These professors all teach introductory physics and use one of a number of popular introductory physics texts. The problems for publication were carefully selected by the professors to cover the fundamental concepts introduced in these standard texts. In general, the problems are just as they appeared in an actual course examination. Furthermore, the solution is that prepared by the professor who wrote the exam problem. Thus this book is a combination of authentic examination problems, together with the professors' own solutions.

No attempt has been made to impose a uniform format on problem solutions. Thus the student will encounter diverse problem solving styles, notation, and conventions for significant figures. Indeed, exposure to this kind of diversity may be considered one of the extra benefits derived from using a book such as this. Athough each contributor has been careful to eliminate errors, perfection is difficult to attain. If you find any errors, a note, mailed to the Engineering Press address, would be appreciated. I hope that this material will help you improve both your understanding of physics and your exam scores. If you find this volume is an aid, you may wish to examine the second and third volumes of the *Physics Exam Files: Volume II, Heat, Light and Sound* and *Volume III, Electricity and Magnetism.*

Arthur B. Western
Editor

EXAM FILES

Professors around the country have opened their exam files and revealed their examination problems and solutions. These are actual exam problems with the complete solutions prepared by the same professors who wrote the problems. EXAM FILES are currently available for these topics:

Circuit Analysis
Dynamics
Engineering Economy
Fluid Mechanics
Materials Science
Mechanics of Materials
Physics I   Mechanics
Physics II Heat, Light and Sound
Physics III Electricity and Magnetism
Probability and Statistics
Statics
Thermodynamics

The EXAM FILE series also includes three engineering license review books:

Engineer-In-Training Exam File
Civil Engineering License Exam File
Mechanical Engineering License Exam File

Other EXAM FILES planned for release in the near future are:

Chemistry
Calculus I
Calculus II
Calculus III

For a description of all available EXAM FILES, or to order them, ask at your college or technical bookstore, or write to:

Engineering Press, Inc.
P.O. Box 1
San Jose, California 95103-0001

1
VECTORS

ADDITION AND SUBTRACTION

━━ 1-1

Consider two displacements, one of magnitude 3 meters and another of magnitude 4 meters. Show diagrammatically how the displacement vectors may be combined to get a resultant displacement of magnitude

   (i) 7 meters   (ii) 1 meter   (iii) 5 meters

****

Convention for directions ⟹

           N
           ↑
  W ────┼──── E
           ↓
           S

We are given two vectors: $\vec{A}$, $\vec{B}$

   ─────────────→ $\vec{A}$
       A = 4 m

   ──────────→ $\vec{B}$
       B = 3 m

We shall use 'end' to 'tip' arrangement of vectors to determine the resultant vector $\vec{A} + \vec{B}$

(i) We want $|\vec{A}+\vec{B}| = 7$

$$\xrightarrow{} \vec{A} \xrightarrow{} \vec{B}$$

$$\xrightarrow{} \vec{A}+\vec{B}$$

$|\vec{A}+\vec{B}| = 7\text{ m}$

(ii) We want $|\vec{A}+(-\vec{B})| = 1\text{ m}$

$\vec{B} \xleftarrow{} \quad \xrightarrow{} \vec{A} \quad \xrightarrow{} \vec{A}-\vec{B}$

$|\vec{A}-\vec{B}| = 1\text{ m}$

(iii) We want $|\vec{A}+\vec{B}| = 5\text{ m}$

By Pythagoras Theorem

$|\vec{A}+\vec{B}| = (A^2 + B^2)^{1/2}$
$\quad = \sqrt{(4m)^2 + (3m)^2}$
$\quad = 5\text{ m}$

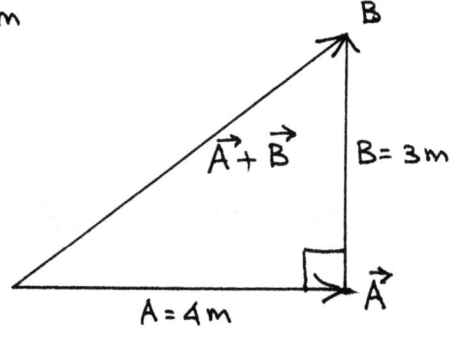

1-2

Given the two vectors $\vec{A} = 2\hat{i} + \hat{j} + 3\hat{k}$ and $\vec{B} = 7\hat{i} - 8\hat{j} + 9\hat{k}$, find the component of $\vec{A}$ in the direction of $\vec{B}$.

****

The angle between $\vec{A}$ and $\vec{B}$ is found from the relation
$$\vec{A}\cdot\vec{B} = AB\cos\theta,$$
where
$$\vec{A}\cdot\vec{B} = A_x B_x + A_y B_y + A_z B_z$$
$$= (2)(7) - (1)(8) + (3)(9) = 33$$
and
$$AB = \sqrt{2^2 + 1^2 + 3^2}\sqrt{7^2 + 8^2 + 9^2} = \sqrt{(14)(194)}.$$
$$\therefore \theta = \cos^{-1}\frac{33}{\sqrt{(14)(194)}} = 50.7°.$$

The component of $\vec{A}$ in the direction of $\vec{B}$ is then,
$$A\cos\theta = \sqrt{14}\cos 50.7° = 2.37 \leftarrow$$

COMPONENTS AND RESULTANTS

1-3

VECTORS A AND B ARE GIVEN IN UNIT VECTOR NOTATION, WHERE THE i AND j ARE UNIT VECTORS IN THE X AND Y DIRECTIONS, RESPECTIVELY.

$\bar{A} = 13i + 27j \qquad \bar{B} = 4i - 14j$

A) FIND THE RESULTING VECTORS OF A + B AND A - B IN UNIT VECTOR NOTATION.

B) FIND A+B AND A-B IN TERMS OF MAGNITUDE AND DIRECTION WITH RESPECT TO THE X DIRECTION.

****

A) $\bar{C} = \bar{A} + \bar{B} = 13i + 27j + 4i - 14j$
$\qquad = 17i + 13j$

$\bar{D} = \bar{A} - \bar{B} = 13i + 27j - (4i - 14j)$
$\qquad = 9i + 41j$

B) Since i and j are perpendicular unit vectors, the magnitude can be found by the Pythagorean Theorem.

$C = |\bar{A}+\bar{B}| = (17^2 + 13^2)^{1/2} = (289 + 169)^{1/2}$
$\qquad = 21.4$

$D = |\bar{A}-\bar{B}| = (9^2 + 41^2)^{1/2} = (81 + 1681)^{1/2}$
$\qquad = 42$

$\theta_C = \tan^{-1}(13/17) = 37.4° \quad (.65 \text{ Rad.})$

$\theta_D = \tan^{-1}(41/9) = 77.6° \quad (1.35 \text{ Rad.})$

1-4

Two tracking stations A and B sight a satellite. Station A reports the position of the satellite to be in the east 45° above the line joining A and B. Station B which is 600 km west of A sights the satellite 20° above the line joining A and B. What is the height of the satellite above the line that goes through A and B?

****

Describing problem:

(We will neglect the curvature of the earth)

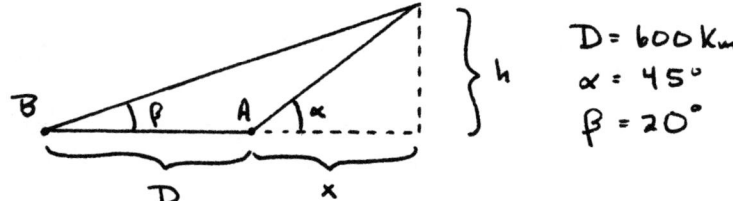

$D = 600 \text{ km}$
$\alpha = 45°$
$\beta = 20°$

Setting up eqs.

$$\tan \alpha = \frac{h}{x} \implies x = \frac{h}{\tan \alpha}$$

$$\tan \beta = \frac{h}{D+x} = \frac{h}{D + \left(\frac{h}{\tan \alpha}\right)}$$

$$\implies \tan \beta \left(D + \frac{h}{\tan \alpha}\right) = h$$

$$\implies D \tan \beta = h - h \frac{\tan \beta}{\tan \alpha} = h\left(1 - \frac{\tan \beta}{\tan \alpha}\right)$$

$$\implies h = D \frac{\tan \beta}{1 - \frac{\tan \beta}{\tan \alpha}}$$

Substituting numbers:

$$h = (600) \frac{\tan(20°)}{1 - \frac{\tan(20°)}{\tan(45°)}} = \underline{\underline{343 \text{ km}}}$$

THE DOT AND CROSS PRODUCTS

1-5

Find the angle between vectors $\mathbb{A}$ and $\mathbb{C}$ given below.

$$\mathbb{A} = 1\hat{i} - 2\hat{j} - 2\hat{k}$$
$$\mathbb{C} = 3\hat{i} - 4\hat{j}$$

****

Use: $\quad \mathbb{A} \cdot \mathbb{C} = |\mathbb{A}||\mathbb{C}| \cos \theta$

Where: $\quad \mathbb{A} \cdot \mathbb{C} = A_x C_x + A_y C_y + A_z C_z$
$$= (1)(3) + (-2)(-4) + (-2)(0) = 11$$

$$|\mathbb{A}| = \sqrt{A_x^2 + A_y^2 + A_z^2} = \sqrt{1^2 + (-2)^2 + (-2)^2} = 3$$

$$|\mathbb{B}| = \sqrt{B_x^2 + B_y^2 + B_z^2} = \sqrt{3^2 + (-4)^2 + 0^2} = 5$$

Solving: $\quad \cos \theta = \dfrac{\mathbb{A} \cdot \mathbb{C}}{|\mathbb{A}||\mathbb{C}|} = \dfrac{11}{(3)(5)} = \dfrac{11}{15} = 0.73\overline{3}$

$$\theta = \cos^{-1}(0.733) = \underline{\underline{42.8°}}$$

1-6

What are the dot and cross products of, and the angle between the following two vectors?

$$\vec{A} = 3\vec{i} + 4\vec{j},$$
$$\vec{B} = 4\vec{j} + 3\vec{k}.$$

***

The dot product is a scalar:

$$\vec{A} \cdot \vec{B} = A_x B_x + A_y B_y + A_z B_z = 3 \times 0 + 4 \times 4 + 0 \times 3 = 16$$

The cross product is a vector:

$$\vec{A} \times \vec{B} = \begin{vmatrix} \vec{i} & \vec{j} & \vec{k} \\ 3 & 4 & 0 \\ 0 & 4 & 3 \end{vmatrix} = \vec{i}(4 \times 3 - 0 \times 4) + \vec{j}(0 \times 0 - 3 \times 3) + \vec{k}(3 \times 4 - 4 \times 0)$$

$$= 12\vec{i} - 9\vec{j} + 12\vec{k}$$

The angle between $\vec{A}$ and $\vec{B}$ is obtained from the dot product:

$$\vec{A} \cdot \vec{B} = |\vec{A}| \cdot |\vec{B}| \cos\theta = 16 \text{ (from above)}$$

$$|\vec{A}| = \sqrt{3^2 + 4^2 + 0} = 5$$

$$|\vec{B}| = \sqrt{0 + 4^2 + 3^2} = 5$$

$$\therefore \cos\theta = \frac{16}{25} = .64 \implies \theta = 50.2°$$

1-7

In each of the illustrated cases, state the magnitude of the dot product and the magnitude and direction of the cross product of the given vectors. Vector A is 4 units long and vector B is 3 units long. A dot represents a vector coming out of the paper and a cross represents a vector going into the paper.

(a) [diagram: vectors A and B with 120° angle between them]
(b) × ←   A   B
(c) •   ×   A   B
(d) ↑A   •B

****

(a) $\vec{A} \cdot \vec{B} = |\vec{A}||\vec{B}| \cos 120° = (4)(3)(-.5) = -6$

$|\vec{A} \times \vec{B}| = |\vec{A}||\vec{B}| \sin 120° = (4)(3)(.866) = 10.4$

Direction of $\vec{A} \times \vec{B}$ is out of the paper (as determined by the right-hand rule).

(b) $\vec{A} \cdot \vec{B} = |\vec{A}||\vec{B}| \cos 90° = 0$

$|\vec{A} \times \vec{B}| = |\vec{A}||\vec{B}| \sin 90° = (4)(3)(1) = 12$

Direction of $\vec{A} \times \vec{B}$ is ↑

(c) $\vec{A} \cdot \vec{B} = |\vec{A}||\vec{B}| \cos 180° = (4)(3)(-1) = -12$

$|\vec{A} \times \vec{B}| = |\vec{A}||\vec{B}| \sin 180° = 0$

Direction of $\vec{A} \times \vec{B}$ is undefined.

(d) $\vec{A} \cdot \vec{B} = |\vec{A}||\vec{B}| \cos 90° = 0$

$|\vec{A} \times \vec{B}| = |\vec{A}||\vec{B}| \sin 90° = (4)(3)(1) = 12$

Direction of $\vec{A} \times \vec{B}$ is →

1-8

The position of a particle is given by the following polynomial expression:

$$\vec{r}(t) = (At^4 + Bt^3)i + (Ct^2 + Dt)j + (E)k$$

where A, B, C, D, and E are arbitrary constants having units of m/sec^4, m/sec^3, m/sec^2, m/sec, and m, respectively. The symbols, i, j, and k denote the unit vectors in the x, y, and z directions. If m is the mass of the particle, find:

(a) $\vec{L} = \vec{r} \times m(d\vec{r}/dt)$; at the time, t = 2 sec.
(b) $\vec{\tau} = \vec{r} \times m(d^2\vec{r}/dt^2)$; at the same time, t = 2 sec.
(c) $W = m(d^2\vec{r}/dt^2) \cdot \vec{r}$; at the time, t = 1 sec.

****

Note: All differentiations must be carried out before any time substitutions or vector operations are performed.

(a) $\vec{L} = [(At^4+Bt^3)\hat{i} + (Ct^2+Dt)\hat{j} + (E)\hat{k}]$
$\times m[(4At^3+3Bt^2)\hat{i} + (2Ct+D)\hat{j}]$

$= [(16A+8B)\hat{i} + (4C+2D)\hat{j} + (E)\hat{k}] \times m[(32A+12B)\hat{i} + (4C+D)\hat{j}]$

$= \begin{vmatrix} \hat{i} & \hat{j} & \hat{k} \\ (16A+8B) & (4C+2D) & E \\ m(32A+12B) & m(4C+D) & 0 \end{vmatrix}$

$= \hat{i}[-Em(4C+D)] + \hat{j}[Em(32A+12B)] + \hat{k}[(16A+8B) \cdot m(4C+D) - m(32A+12B)(4C+2D)]$

$= m[\hat{i}[-E(4C+D)] + \hat{j}[E(32A+12B)] + \hat{k}[(16A+8B)(4C+D) - (32A+12B)(4C+2D)]]$

(b) $\vec{\tau} = [(At^4+Bt^3)\hat{i} + (Ct^2+Dt)\hat{j} + (E)\hat{k}]$
$\times m[(12At^2+6Bt)\hat{i} + (2C)\hat{j}]$

$= [(16A+8B)\hat{i} + (4C+2D)\hat{j} + (E)\hat{k}] \times m[(48A+12B)\hat{i} + (2C)\hat{j}]$

$= \begin{vmatrix} \hat{i} & \hat{j} & \hat{k} \\ (16A+8B) & (4C+2D) & E \\ m(48A+12B) & m(2C) & 0 \end{vmatrix}$

$$= \hat{i}[-m(2EC)] + \hat{j}[mE(48A+12B)] + \hat{k}[m(16A+8B)(2C) - m(48A+12B)(4C+2D)]$$

$$= m[\hat{i}[2EC] + \hat{j}[E(48A+12B)] + \hat{k}[(16A+8B)(2C) - (48A+12B)(4C+2D)]]$$

(c) $W = m[(12At^2+6Bt)\hat{i} + (2C)\hat{j}] \cdot [(At^4+Bt^3)\hat{i} + (Ct^2+Dt)\hat{j} + (E)\hat{k}]$

$$= m[(12A+6B)\hat{i} + (2C)\hat{j}] \cdot [(A+B)\hat{i} + (C+D)\hat{j} + (E)\hat{k}]$$

$$= m[(12A+6B)(A+B) + (2C)(C+D) + 0]$$

$$= m[12A^2 + 12AB + 6AB + 6B^2 + 2C^2 + 2CD]$$

$$= 2m[6A^2 + 9AB + 3B^2 + C^2 + CD]$$

1-9

Two vectors are $\vec{A} = 2\hat{i}$ and $\vec{B} = 3\hat{i} - 4\hat{j}$, where $\hat{i}$, $\hat{j}$, and $\hat{k}$ are unit vectors in the x, y, and z directions respectively. Find the magnitude and direction of $\vec{A} \times \vec{B}$.

****

$\vec{A} \times \vec{B} = (2\hat{i}) \times (3\hat{i} - 4\hat{j}) = 6\underbrace{\hat{i} \times \hat{i}}_{=0} - 8\underbrace{\hat{i} \times \hat{j}}_{\hat{k}} = \underline{\underline{-8\hat{k}}}$

Alternatively, $|\vec{A} \times \vec{B}| = |\vec{A}||\vec{B}|\sin\theta = (2)(5)\sin[\sin^{-1}(\frac{4}{5})] = (2)(5)(\frac{4}{5}) = \underline{\underline{8}}$

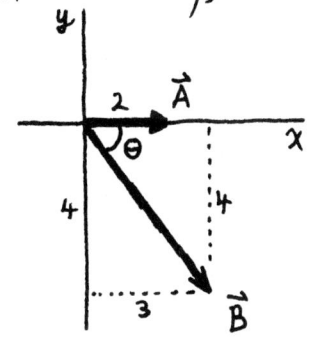

and by the right-hand rule the direction of $\vec{A} \times \vec{B}$ is $\perp$ <u>into the page</u>, i.e. the negative z direction.

1-10

For the vectors $\vec{A}$, $\vec{B}$, $\vec{C}$, shown below, calculate
a) $\vec{A} + \vec{B} + \vec{C}$, b) $|\vec{A} + \vec{B} + \vec{C}|$, c) $\vec{A} \cdot \vec{B}$, and d) $\vec{C} \times \vec{A}$.

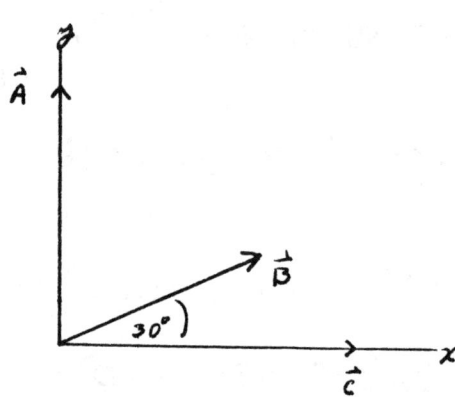

Note: $|\vec{A}| = 6$, $|\vec{B}| = 5$, $|\vec{C}| = 7$.

****

Let $\hat{\imath}, \hat{\jmath}, \hat{k}$ be unit vector along the x-y-z axes respectively. Then

$$\vec{A} = 6\hat{\jmath}\;;\quad \vec{B} = 5\cos 30\,\hat{\imath} + 5\sin 30\,\hat{\jmath}\;;\quad \vec{C} = 7\hat{\imath}$$

a) $\vec{A} + \vec{B} + \vec{C} = 6\hat{\jmath} + 4.33\hat{\imath} + 2.5\hat{\jmath} + 7\hat{\imath}$
$\qquad = 11.33\hat{\imath} + 8.5\hat{\jmath}$

b) $|\vec{A} + \vec{B} + \vec{C}| = \sqrt{(11.33)^2 + (8.5)^2} = 14.16$

c) $\vec{A} \cdot \vec{B} = |A||B|\cos 60$ $\qquad$ —OR— $\vec{A}\cdot\vec{B} = A_x B_x + A_y B_y$
$\qquad = 6 \cdot \left(\sqrt{(4.33)^2 + (2.5)^2}\right)(.5)$ $\qquad\qquad = 0 + 6(2.5)$
$\qquad = 15.$ $\qquad\qquad\qquad\qquad\qquad\qquad\qquad = 15.$

d) $\vec{C} \times \vec{A} = |C||A|\sin 90\,\hat{k}$ $\qquad$ —OR— $\vec{C} \times \vec{A} = (C_y A_z - C_z A_y)\hat{\imath} +$
$\qquad = 7 \cdot 6\,\hat{k}$ $\qquad\qquad\qquad\qquad\qquad (C_z A_x - C_x A_z)\hat{\jmath} +$
$\qquad = 42\,\hat{k}$ $\qquad\qquad\qquad\qquad\qquad (C_x A_y - C_y A_x)\hat{k}$
$\qquad\qquad\qquad\qquad\qquad\qquad\qquad\qquad\;\; = 42\,\hat{k}$

You are given the vectors $\underline{A} = 3\hat{i} + 4\hat{j}$ and $\underline{B} = -3\hat{i} + 3\hat{j}$. Determine the magnitude and direction (as an angle measured from the +x-axis) of $\underline{A}$, of $\underline{B}$, of $\underline{A+B}$, and of $\underline{A-B}$. Also find the dot product $\underline{A} \cdot \underline{B}$.
Draw a careful sketch of the vectors on the grid at the right.

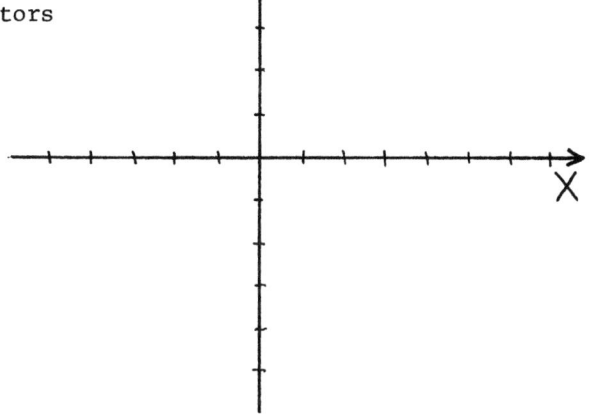

****

$|A| = \sqrt{A_x^2 + A_y^2} = \sqrt{9+16} = 5$; $\theta_A = \arctan\frac{4}{3} = 53.1°$

$|B| = \sqrt{9+9} = 4.24$; $\theta_B = \arctan(-1) = 135°$

$\vec{A+B} = 0 + 7\hat{j}$, so $|A+B| = 7$; $\theta_{A+B} = 90°$

$\vec{A-B} = 6\hat{i} + 1\hat{j}$, so $|A-B| = \sqrt{36+1} = 6.1$; $\theta_{A-B} = \arctan\frac{1}{6} = 9.5°$

$\vec{A} \cdot \vec{B} = A_x B_x + A_y B_y = 3(-3) + 4(3) = +3$.

Check with other method: $\vec{A} \cdot \vec{B} = |A||B|\cos\phi = 5(4.24)\cos(135-53.1)$
$= 3.0$ ✓

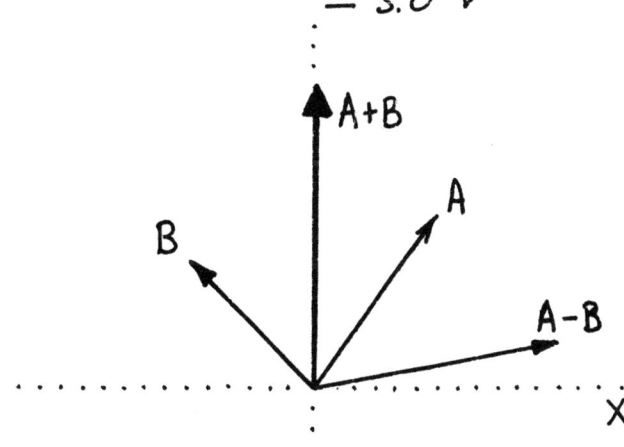

1-12

Vectors $\vec{A}$ and $\vec{B}$ have components $(A_x, A_y, A_z)=(3,2,0)$ and $(B_x, B_y, B_z)=(-1,1,-2)$. Find the vector $\vec{A}-\vec{B}$ in component form; find its length, and the angle it makes with the x-y plane.

****

$$\vec{A} - \vec{B} = (A_x - B_x, A_y - B_y, A_z - B_z) = (4, 1, 2)$$

$$|\vec{A} - \vec{B}| = \sqrt{4^2 + 1^2 + 2^2} = 4.58$$

Let θ_z be the angle $\vec{A} - \vec{B}$ makes with the z-axis.

Then $\cos \theta_z = \dfrac{\hat{k} \cdot (\vec{A} - \vec{B})}{|\vec{A} - \vec{B}|} = \dfrac{2}{4.58}$ $\therefore \theta_z = 64.12°$

$\therefore$ Angle made with x-y plane = $90° - \theta_z = 25.88°$

1-13

Find the angles between the vector $\vec{A}$ and each of the three axes.

$\vec{A} = 2i + 3j + 6k$

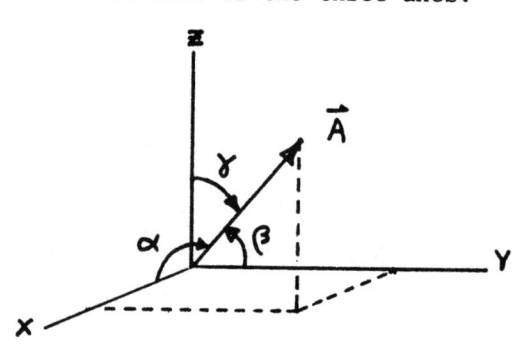

****

$\vec{A} \cdot \vec{i} = A(1) \cos \alpha$ $A = \sqrt{2^2 + 3^2 + 6^2} = 7$

$\cos \alpha = \dfrac{\vec{A} \cdot \vec{i}}{A} = \dfrac{2}{7}$ $\alpha = 73.4°$

$\cos \beta = \dfrac{\vec{A} \cdot \vec{j}}{A} = \dfrac{3}{7}$ $\beta = 64.6°$

$\cos \gamma = \dfrac{\vec{A} \cdot \vec{k}}{A} = \dfrac{6}{7}$ $\gamma = 31.0°$

2
VELOCITY AND ACCELERATION

AVERAGE VELOCITY

■■2-1

An aircraft flies a straight route at constant speed. Compute the speed from this radar data: At 9 AM, the plane is 100 miles from the radar, at azimuth 30 degrees west of north. One hour and 15 minutes later, the plane is 200 miles from the radar, in the direction due east.

✶✶

data
$$\begin{cases} \vec{r_2} = i\,200 + j\,0 \\ \vec{r_1} = -i\,100 \sin 30° + j\,100 \cos 30° \\ \phantom{\vec{r_1}} = -i\,50 + j\,86.6 \end{cases}$$

$$\vec{\Delta r} = \vec{r_2} - \vec{r_1}$$
$$= i(200+50) + j(0-86.6)$$
$$= i\,250 - j\,86.6 \text{ miles}$$

$$\Delta r = (250^2 + 86.6^2)^{\frac{1}{2}} = 265 \text{ miles}$$

$$v = \frac{\Delta r}{\Delta t} = \frac{265 \text{ mi}}{1.25 \text{ hr}} = \underline{\underline{212 \text{ mi/hr}}}$$

2-2

To qualify for a race, a bicyclist needs to achieve an average speed of 60 ft/sec in her trial heats. In one such heat, the bicyclist covers the first half of the 600-foot track at an average speed of 45 ft/sec. At what average speed must she cover the second half of the track in order to qualify?

***

The first thing NOT to do is to conclude that the average speed is the same as the average of the speeds! That is to say, well, she covered the first half of the track at 45 ft/sec, and she needs to average 60 ft/sec, so if she does the second half at 75 ft/sec, she'll achieve the requisite average speed (45 + 75 = 120, ÷2 = 60 ft/sec). THIS IS INCORRECT!! The average speed is found by dividing the total distance by the total time. The total distance is known — that's 600 feet. To achieve an average speed of 60 ft/sec over this distance, the bicyclist must cover this distance in a time

$$t = d/v = 600\text{ft}/60\text{ft/sec} = 10 \text{ seconds}$$

How much time has she used so far? She covered half the track (300 ft) at an average speed of 45 ft/sec

$$\therefore \text{she's used } t_1 = d_1/v_1 = 300\text{ft}/45\text{ft/sec} = \frac{20}{3} \text{ sec.}$$

This means she must complete the second half of the track in $(10 - \frac{20}{3}) = 10/3$ sec. She must cover half the track (300 ft) in $10/3$ sec $\Rightarrow$ her average speed for part 2 $= v_2 = d_2/t_2 = \frac{300\text{ft}}{10/3 \text{ sec}} = \boxed{90 \text{ ft/sec}}$

INSTANTANEOUS VELOCITY

2-3

Consider a car standing at a red light. When the light turns green, the car's distance from the light is seen to change as a function of time as $x = 6t + 2t^2$ meters.

Calculate a) its average velocity during the first three seconds, and
b) its instantaneous velocity at t=3 sec.

****

a) $\vec{v}_{ave.} = \dfrac{\Delta x}{\Delta t} \hat{\imath}$

$= \dfrac{x(t=3) - x(t=0)}{3} \hat{\imath}$

$= \dfrac{6 \cdot 3 + 2 \cdot 3^2 - 0}{3} \hat{\imath}$

$= 12 \hat{\imath}$ m./sec.

b) $\vec{v}_{inst.}(t=3) = \left. \dfrac{dx}{dt} \hat{\imath} \right|_{t=3 \text{ sec.}}$

$= \left. (6 + 4t)\hat{\imath} \right|_{t=3 \text{ sec.}}$

$= (6 + 12)\hat{\imath}$

$= 18 \hat{\imath}$ m./sec.

2-4

The position of a particle as it moves on the x-axis is shown in the tabular data at the right. From these data determine the average speed of the particle during the interval 0-2 sec and 0-5 sec.
Is this particle subject to a constant acceleration?

Position (meters)	Time (sec)
1.0	0.0
3.0	1.0
4.0	2.0
3.0	3.0
1.0	4.0
-3.0	5.0

***

Average speed is defined by $V_{ave} = \dfrac{X_f - X_i}{t_f - t_i}$

for the period 0-2 sec $X_f = 4m$ and $X_i = 1m$

$$V_{ave} = \frac{4-1}{2-0} \frac{(meter)}{(sec)} = 1.5 \text{ m/sec}$$

for the period 0-5 sec $X_f = -3m$ and $X_i = 1m$

$$V_{ave} = \frac{-3.0 - 1.0}{5 - 0} = \frac{-4}{5} = -.8 \text{ m/sec}$$

If the particle's motion is under constant acceleration then

$$X_f = X_i + V_0 t + \tfrac{1}{2} a t^2$$

Get two equations with V_0 and a as unknowns from the initial position when $t=0$ and using X_f as the position when $t=1$ sec and $t=2$ sec

$t = 1 \text{ sec} \Rightarrow 3 = 1 + V_0 + \tfrac{1}{2} a$

$t = 2 \text{ sec} \Rightarrow 4 = 1 + 2V_0 + 2a$

Solve for V_0 and a get $V_0 = 2.5$ m/sec, $a = -1$ m/sec^2

Using the obtained values for a and V_0 see if the correct value of x at $t=4$ or $t=5$ sec is given

$t = 4$ sec

$$X_f = 1 + 2.5 \times 4 - \tfrac{1}{2} \times 4^2 = 3m \text{ — does not agree}$$

∴ not constant acceleration

2-5

A particle moves along the x-axis according to the relation $x = (2/3)t^3 - 3t^2 - 8t + 1$, where x is in cm, t is in sec: (a) When is the particle moving in the negative x-direction? (b) Sketch v(t) and a(t).

***

(a) The velocity of the particle v is,
$$v = \dot{x} = 2t^2 - 6t - 8 = 2(t+1)(t-4),$$
which has negative values for $-1 < t < 4$. Therefor, the particle is moving in the negative x-direction during $-1 < t < 4$.

(b)

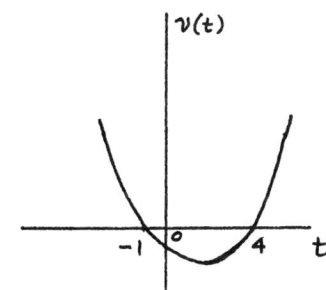
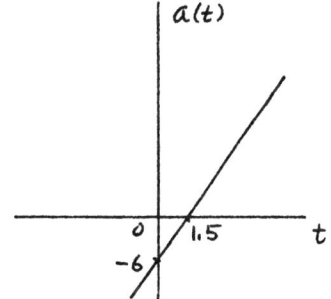

AVERAGE ACCELERATION

2-6

A rock is tossed upward at 32 feet/second. How high does it go until its upward speed decreases to zero? This upward trip of the rock requires a certain amount of time for the speed to decrease from 32 feet/second to 0.0 feet/second. How long?

***

$$v^2 = v_0^2 + 2as \qquad s = vt - \tfrac{1}{2}at^2$$
$$v^2 = v_0^2 + 2(-g)h \qquad h = 0 - \tfrac{1}{2}(-g)t^2$$
$$v^2 = v_0^2 - 2gh \qquad h = \tfrac{1}{2}gt^2$$

$$v^2 + 2gh = v_0^2$$
$$2gh = v_0^2 - v^2$$
$$h = \frac{v_0^2 - v^2}{2g}$$
$$h = \frac{(32 \text{ feet/sec})^2 - (0)^2}{2(32 \text{ feet/sec}^2)}$$
$$\boxed{h = 16 \text{ feet}}$$

$$t^2 = \frac{2h}{g}$$
$$t^2 = \frac{2(16 \text{ feet})}{32 \text{ feet/sec}^2}$$
$$t^2 = 1.0 \text{ sec}^2$$
$$\boxed{t = 1.0 \text{ sec}}$$

INSTANTANEOUS ACCELERATION

2-7

The velocity of a particle moving along the x-axis is described by the following function:

$$v = (3 \text{ m/s}) + (4 \text{ m/s}^4) t^3$$

Find (a) the average acceleration between t=1 sec and t=3 sec, and (b) the instantaneous acceleration at t=2 sec.

****

(a) $a_{av} = \frac{\Delta v}{\Delta t} = \frac{v_{final} - v_{initial}}{t_{final} - t_{initial}} = \frac{111 \text{ m/s} - 7 \text{ m/s}}{3\text{s} - 1\text{s}} = 52 \text{ m/s}^2$

(b) $a_{inst} = \frac{dv}{dt}\Big|_{t=2\text{sec}} = \left[3(4 \text{ m/s}^4) t^2\right]_{t=2\text{sec}} = 48 \text{ m/s}^2$

2-8

A 5-kg object moves along the x-axis. Its position, in meters, as a function of the time, in seconds, is

$$x = at - bt^3$$

where $a = 3$ and $b = 2$.

(a) What must be the units for the constant, b?
(b) Find the average velocity of this object in the time interval $t = 1$ to $t = 3$ seconds.
(c) Find the instantaneous velocity at $t = 2$ seconds.
(d) Find the acceleration of the object as a function of time, in seconds.

* *

(a) "bt^3" must give "meter" when units are cancelled, so
(units of b) × (second)^3 = meter
∴ units of b = m/s^3

(b) $\overline{v_x}$ (Def.) = $\frac{\Delta x}{\Delta t}$ = $\frac{x(3) - x(1)}{3 - 1}$ and $x(t) = 3t - 2t^3$
$x(3) = 3 \cdot 3 - 2 \cdot 3^3 = -45\, m$; $x(1) = 3 \cdot 1 - 2 \cdot 1^3 = +1\, m$
$\overline{v_x} = \frac{-45 - (+1)}{2} = -23\, m/s$, i.e. toward $(-x)$ direction.

(c) v_x (Def.) = $\frac{dx}{dt} = a - 3bt^2 = 3 - 6t^2\, m/s$
At $t = 2$, $v_x(2) = 3 - 6 \cdot 2^2 = -21\, m/s$

Note that the average velocity is NOT the instantaneous velocity at the midpoint of the time interval, NOR is it $\frac{1}{2}[v_x(3) + v_x(1)] = \frac{1}{2}[-51 - 3] = -27\, m/s$, which is correct only for <u>constant</u> acceleration. This is NOT so here, because

(d) a_x (Def.) = $\frac{dv_x}{dt} = -6bt = -12t\, m/s^2$, and this varies with time.

The 5 kg mass of the object is not a part of the <u>kinematical</u> description of the motion. It is needed to determine the <u>forces</u> involved; that is, the <u>dynamics</u> of the motion.

2-9

A particle in one-dimensional motion has speed v(m/s) at any instant t given by v=bt^2+ct+e, where b, c, and e are constants having the numerical values b=3, c=4, and e=6.
(a) Find the initial speed.
(b) Find the speed when 3 seconds have passed.
(c) Find the expression for a(t), the instantaneous acceleration in terms of time t, from the above expression for v(t).
(d) From (c) what is the acceleration when 4 seconds have elapsed?
(e) Is the problem being treated an example of constant acceleration? (Show)
(f) Find an expression for the displacement x(t) using the given equation for v(t).

****

(a) $v(0) = e = 6$ m/s

(b) $v(3) = 3(3)^2 + 4(3) + 6 = 45$ m/s

(c) $a(t) = \dfrac{dv(t)}{dt} = 2bt + c = 6t + 4$ m/s^2

(d) $a(4) = 28$ m/s^2

(e) $a = 2bt + c$ ∴ $a = a(t)$; accel. is not const.

(f) $v = bt^2 + ct + e$

$\dfrac{dx}{dt} = bt^2 + ct + e$

$\displaystyle\int_{x_0}^{x} dx' = b\int_0^t t'^2 dt' + c\int_0^t t' dt' + e\int_0^t dt'$

$x - x_0 = b\dfrac{t^3}{3} + c\dfrac{t^2}{2} + et$

∴ $x = x_0 + t^3 + 2t^2 + 6t$ (m)

2-10

AN ACROBATIC AIRPLANE HAS THE FOLLOWING DISPLACEMENT VECTOR, R, WITH TIME (R IS IN METERS)

$$\bar{R} = 3t^3 i + 4t^2 j + 6t k$$

WHERE i, j and k ARE UNIT VECTORS IN THE X,Y AND Z DIRECTIONS, AND t IS IN SECONDS.

AFTER 10 SECONDS, FIND THE FOLLOWING

A) DISPLACEMENT VECTOR
B) INSTANTANEOUS AND AVERAGE VELOCITY VECTORS
C) INSTANTANEOUS AND AVERAGE ACCELERATION VECTORS

****

A) Letting t=10 seconds in the Displacement formula

$$\bar{R} = 3 \cdot (10)^3 i + 4(10)^2 j + 6t k$$
$$= 3E3 \, i + 4E2 \, j + 6t k$$

B) Instantaneous velocity is found by differentiating the displacement formula with respect to t

$$\bar{V} = \frac{d\bar{R}}{dt} = 9t^2 i + 8t j + 6 k$$

Letting t = 10 seconds

$$\bar{V} = 9E2 \, i + 80 \, j + 6 k$$

Average velocity is found by dividing the change in displacement during the time interval by the time interval

$$V_{ave} = \frac{\Delta \bar{R}}{\Delta t} = \frac{3E3 i + 4E2 j + 60 k - 0}{10} = 3E2 i + 40 j + 6 k$$

C) Intantaneous acceleration if found by differentiating the velocity formula with respect to t

$$\bar{a} = \frac{d\bar{V}}{dt} = 18t \, i + 8 j$$

Letting t = 10 seconds

$$\bar{a} = 180 \, i + 8 j$$

Average acceleration is found by dividing the change in velocity during the time interval by the time interval

$$\bar{a} = \frac{\Delta V}{\Delta t} = \frac{9E2 \, i + 80 j + 6k - 6k}{10}$$
$$= 90 \, i + 8 j$$

2-11

Consider a body whose velocity along the x axis is given as shown below. Calculate a) the average acceleration for the period t = 0 to t = 2, t = 2 to t = 5, and t = 5 to t = 7 secs.
b) the instantaneous acceleration at t = 1 and t = 6 secs.

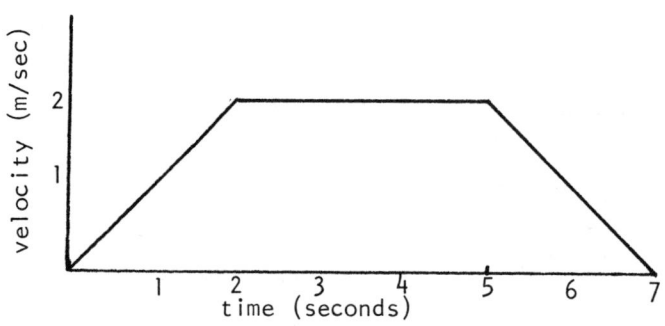

****

a) From the graph's straight line from $t=0$ to $t=2$
$\vec{v} = t\hat{i}$ and since $\vec{a}_{ave} = \frac{\Delta \vec{v}}{\Delta t}$

$$\vec{a}_{ave} = \frac{\vec{v}(t=2) - \vec{v}(t=0)}{2}\hat{i}$$

$$= \frac{2-0}{2}\hat{i}$$

$$= 1\hat{i} \, \frac{m}{s^2}$$

For $t=2$ to $t=5$, $\vec{v} = 2\hat{i}$
$\therefore \vec{a}_{ave} = \frac{\vec{v}(t=5) - \vec{v}(t=2)}{3}$

$$= \frac{2-2}{3}$$

$$= 0$$

For $t=5$ to $t=7$, $\vec{v} = -t\hat{i}$
$\therefore \vec{a}_{ave} = \frac{\vec{v}(t=7) - \vec{v}(t=5)}{2}$

$$= \frac{0-2}{2}\hat{i}$$

$$= -1\hat{i} \, m/s^2$$

b) Since $\vec{v} = t\hat{i}$ from $t=0$ to $t=2$, at $t=1$ sec.
$\vec{a}_{inst} = \frac{d\vec{v}}{dt}\Big|_{t=1s} = 1\hat{i} \, m/s^2$

Since $\vec{v} = -t\hat{i}$ from $t=5$ to $t=7$, at $t=6$ sec.
$\vec{a}_{inst} = \frac{d\vec{v}}{dt}\Big|_{t=6s} = -1\hat{i} \, m/s^2$

2-12

Shown on the graph below is the speed of a particle moving along a straight line which is subject to various constant accelerations. Draw a graph of the acceleration versus time.
If the particle goes through the origin at t = 0, where is it at the end of 4 seconds?

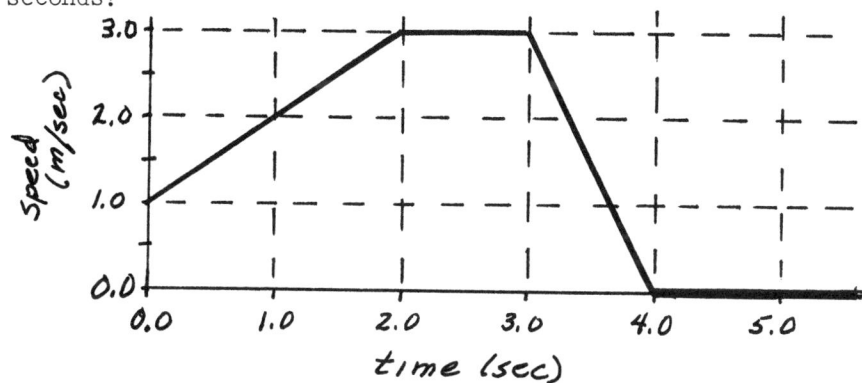

****

Note all accelerations are constant = a_{ave} for period

For the period 0-2 sec

$$a_{ave} = \frac{V(2sec) - V(0sec)}{2 - 0} = \frac{3-1}{2} = 1 \, m/sec^2$$

For the period 2-3 sec

$$\Delta V = 0 \Rightarrow a = 0$$

For the period 3-4 sec

$$a = \frac{V(4) - V(3)}{4-3} = \frac{0-3}{1} = -3 \, m/s^2$$

For the period 4-5 sec $\Delta V = 0 \Rightarrow a = 0$

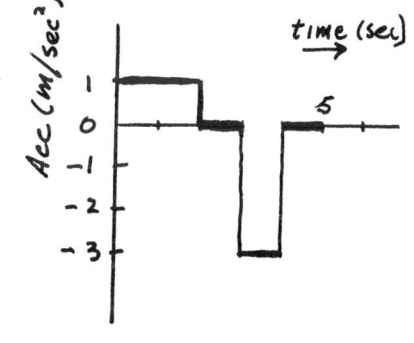

For the distance traveled compute the area under the curve. Determine the area for the segments 0-2 sec, 2-3 sec and 3-4 sec then add them

0-2 Sec – Area = 2×2 = 4 meters   ($V_{average}$ = 2 m/sec)
2-3 Sec – Area = 3 m/sec × 1 sec = 3 meters
3-4 Sec – Triangle area = ½ × 3 × 1 = 1.5 meters

Distance traveled = 4 + 3 + 1.5 = 8.5 m.
particle at origin, x=0, at t=0
∴ x_{final} = 8.5 m

2-13

A particle moves along the x-axis. At time t = 0 it is at x = −5 meters. Its instantaneous velocity in meters per second is given as a function of t in seconds by

$$v = 6t^2 - 8t + 3$$

a) What is its velocity at t = 2 seconds?
b) What is its instantaneous acceleration at t = 3 seconds?
c) What is its average acceleration for the time interval t = 2 to t = 4 seconds?
d) Where is the particle at t = 3 seconds?

****

a) $v(2) = 6 \cdot 4 - 8 \cdot 2 + 3 = 11.0 \text{ m/s}$

b) $a = \dfrac{dv}{dt} = 12t - 8$

at $t = 3s$, $a = 12 \cdot 3 - 8 = 28 \text{ m/s}^2$

c) $\bar{a} = \dfrac{v(4) - v(2)}{2}$

$v(4) = 6 \cdot 4^2 - 8 \cdot 4 + 3 = 67 \text{ m/s}$

$\bar{a} = \dfrac{67 - 11}{2} = 28 \text{ m/s}^2$

d) $x(3) - x(0) = \int_0^3 v \, dt = \int_0^3 (6t^2 - 8t + 3) \, dt$

$x(3) - (-5) = \left(\dfrac{6t^3}{3} - \dfrac{8t^2}{2} + 3t \right) \Big|_0^3$

$x(3) = -5 + (2 \cdot 27 - 4 \cdot 9 + 9) - 0$

$x(3) = 22.0 \text{ m}$

2-14

A particle is moving along the x-axis; the x-component of its velocity is plotted at the right, versus time.

From that, sketch careful but qualitative graphs of its position, $x(t)$, and its acceleration, $A_x(t)$.

Assume that it begins at the origin at $t = 0$.

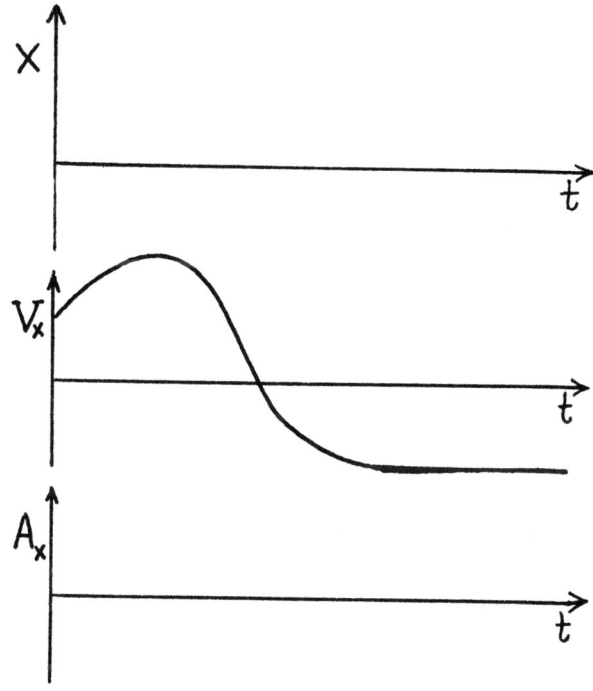

* * *

$$x = \int_0^t V_x(t')\,dt'$$:

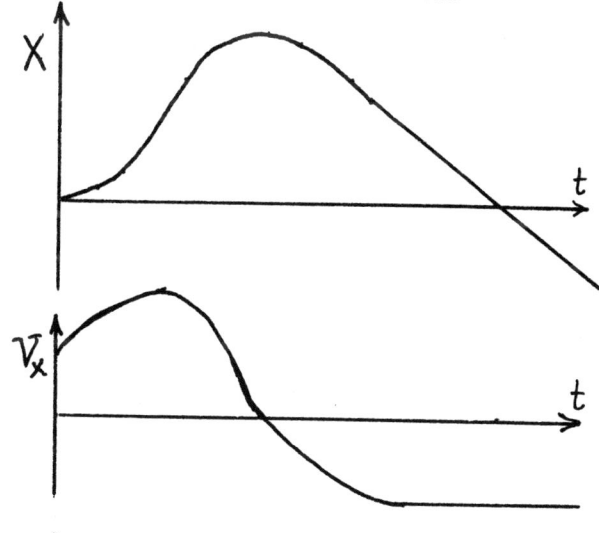

$$A_x = \frac{d}{dt}V_x$$:

Remember:
a) increasing $f(t) \Rightarrow \frac{d}{dt}f > 0$
b) flat $f(t) \Rightarrow \frac{df}{dt} = 0$
c) decreasing $f(t) \Rightarrow \frac{df}{dt} < 0$

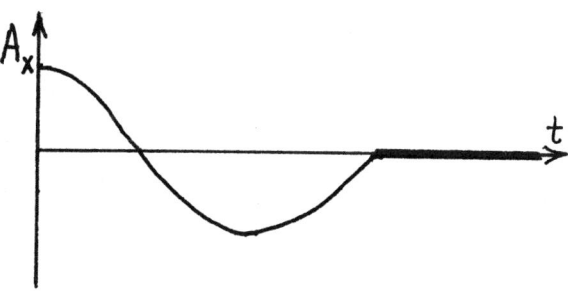

2-15

A particle moves along the x axis according to the equation $x = 6 + 2t - 5t^3$. The position x is in metres, and t is in seconds.

(a) What must the units of the "6" be?
(b) What must the units of the "2" be?
(c) What must the units of the "5" be?
(d) What is the particle's position at t = 1?
(e) What is its velocity at t = 2?
(f) What is its acceleration at t = 1?
(g) Would it be permissable to use the equation $v^2 = v_0^2 + 2a(x - x_0)$ for this particle?
(h) Why or why not?

***

Each term in the equation must have the same units, so the answers to a, b, and c are:

(a) metres   (b) metres/sec   (c) metres/sec^3

(d) $x = 6 + 2(1) - 5(1)^3 = 6 + 2 - 5 = \underline{+3 \text{ metres}}$

(e) $v = \frac{dx}{dt} = 2 - 15t^2 = 2 - 15(2)^2 = 2 - (15)(4) = \underline{\underline{-58 \text{ metres/sec}}}$

(f) $a = \frac{dv}{dt} = -30t = -30(1) = \underline{\underline{-30 \text{ metres/sec}^2}}$

(g) No.   (h) ... because $a = -30t \neq$ constant, but the equation in (g) was developed for the case of <u>constant</u> acceleration.

2-16

A graph of the velocity of an object as a function of time is shown below. (a) Draw an acceleration vs time graph for this object, and don't forget to label and scale the axes. (b) At what time(s) was the object at rest? (c) What was the object's displacement after 5 seconds?

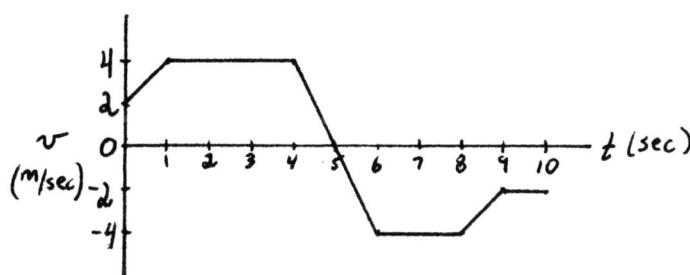

***

(a) The key is realizing that $a = \frac{dv}{dt}$ = the slope of the v vs t graph

(ie) between 4-6 sec
$$a = \frac{\Delta v}{\Delta t} = \frac{-4-4}{2} = -4 \text{ m/sec}^2$$

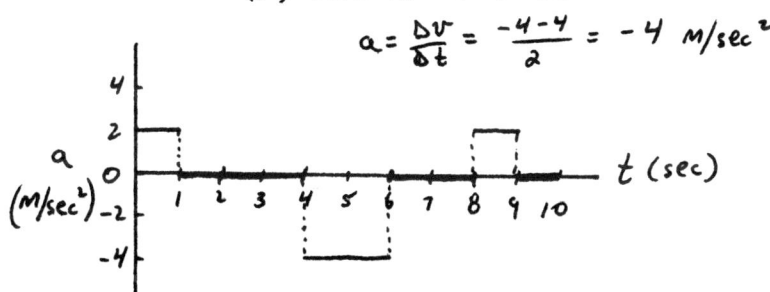

(b) When $v = 0$ the particle is at rest ∴ at $\underline{\underline{t = 5 \text{ sec}}}$

(c) Since $v = \frac{dx}{dt}$

$x = \int v \, dt$ = Area under the v vs t curve

All we do is add the areas for each segment separately

0-1 sec : $A = (2 \times 1) + \frac{1}{2}(2 \times 1) = 3$ m
1-4 sec : $A = 4 \times 3 = 12$ m
4-5 sec : $A = \frac{1}{2}(4 \times 1) = 2$ m

TOTAL Area = $\underline{\underline{17 \text{ m}}}$ = Displacement

2-17

A particle moves in the x-y plane. Its acceleration in the x-direction is $a_x = \sin(t)$ and in the y-direction $a_y = \cos(t)$. If the particle starts at the point x=0, y=1 at t=0 with velocity $v_x = -\cos(t)$, $v_y = 0$ find equations for the x and y components of the motion at time t, and describe the motion using a diagram. Write an expression for the distance from the origin at time t.

****

$a_x = \sin t \Rightarrow v_x = -\cos t + \underset{=0 \text{ from boundary condn.}}{c}$

$\therefore x = -\sin t + \underset{=0 \text{ from initial condn.}}{c'}$

$a_y = \cos t \Rightarrow v_y = \sin t + \underset{=0}{c''}$

$\therefore y = -\cos t + c'''$

At $t=0$: $1 = -1 + c''' \Rightarrow c''' = 2$

$$\therefore (x,y) = (-\sin t, 2 - \cos t)$$

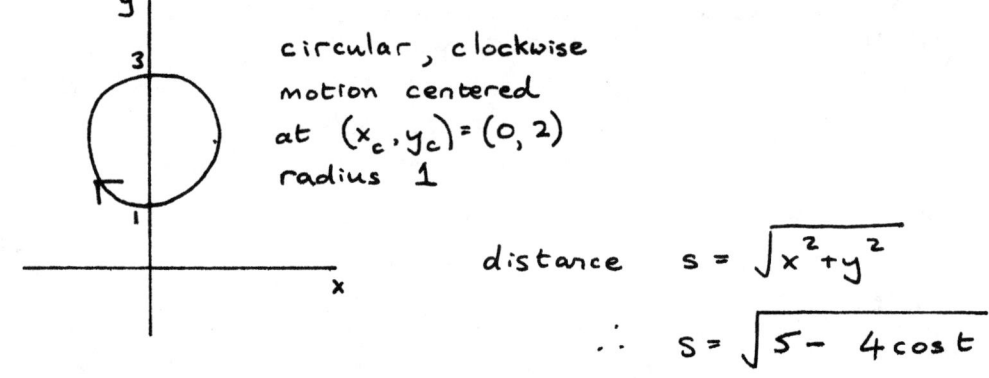

circular, clockwise motion centered at $(x_c, y_c) = (0, 2)$ radius 1

distance $s = \sqrt{x^2 + y^2}$

$\therefore s = \sqrt{5 - 4\cos t}$

Instantaneous Acceleration / 29

2-18

The position of a particle is described by $\vec{r} = (4t^3 - 12t + 9)\hat{i} + (6t + 4)\hat{j}$ meters, where t is in seconds. Use only cartesian coordinates and calculate the following:

a) The instantaneous velocity at t = 0, 1, 2, and 3 seconds.
b) The average velocity between t = 1 and t = 2 seconds.
c) The speed at t = 1 seconds.
d) The instantaneous acceleration at t = 2 seconds.
e) The average acceleration between t = 1 and t = 3 seconds.
f) The time at which the x-component of the position is a relative maximum or minimum.

****

a) $\vec{v} = \dfrac{d\vec{r}}{dt} = (12t^2 - 12)\hat{i} + 6\hat{j}$ m/s

$\vec{v}(0) = -12\hat{i} + 6\hat{j}$, $\vec{v}(1) = 0\hat{i} + 6\hat{j}$ m/s
$\vec{v}(2) = 36\hat{i} + 6\hat{j}$, $\vec{v}(3) = 96\hat{i} + 6\hat{j}$ m/s

b) $\vec{v}_{av} = \dfrac{\Delta \vec{r}}{\Delta t} = \dfrac{\vec{r}(2) - \vec{r}(1)}{2 - 1}$ m/s

$\vec{r}(1) = (4 - 12 + 9)\hat{i} + 10\hat{j} = \hat{i} + 10\hat{j}$ m
$\vec{r}(2) = (32 - 24 + 9)\hat{i} + 16\hat{j} = 17\hat{i} + 16\hat{j}$ m
$\vec{r}(2) - \vec{r}(1) = 16\hat{i} + 6\hat{j}$ m
$\vec{v}_{av} = 16\hat{i} + 6\hat{j}$ m/s

c) Speed $= |\vec{v}(t)|$ so $|\vec{v}(1)| = |0\hat{i} + 6\hat{j}| = 6$ m/s

d) $\vec{a} = \dfrac{d\vec{v}}{dt} = 24t\hat{i} + 0\hat{j}$ m/sec^2

$\vec{a}(2) = 48\hat{i}$ m/sec^2

e) $\vec{a}_{av} = \dfrac{\Delta \vec{v}}{\Delta t} = \dfrac{\vec{v}(3) - \vec{v}(0)}{3 - 0} = \dfrac{(96\hat{i} + 6\hat{j}) - (-12\hat{i} + 6\hat{j})}{3 \text{ sec}}$ m/s

$= \dfrac{108\hat{i} \text{ m/s}}{3 s} = 36\hat{i}$ m/sec^2

f) $\dfrac{dr_x}{dt} = v_x = 12t^2 - 12$ m/sec. This will be zero at a relative maximum or minimum. $t^2 = 1$, $t = 1$ sec. A plot of r_x shows a minimum at this time.

2-19

The position of a particle moving in one dimension as a function of time is given by $x = 6 + 3t^2 - 2t^3$ where x is in m and t is in seconds.
(a) Find the instantaneous velocity at t = 1 s and t = 3 s. (b) Find the instantaneous acceleration at t = 2 s. (c) Find the average velocity from t = 1 s to t = 3 s.

***

a) $v_x = \dfrac{dx}{dt} = 6t - 6t^2$

$v_x(1s) = 6 - 6 = 0 \text{ m/s}$

$v_x(3s) = 6(3) - 6(3)^2 = 18 - 54 = -36 \text{ m/s}$

b) $a_x = \dfrac{dv}{dt} = 6 - 12t$

$a(2s) = 6 - 12(2) = -18 \text{ m/s}^2$

c) $x(1) = 6 + 3(1)^2 - 2(1)^3 = 7 \text{ m}$

$x(3) = 6 + 3(3)^2 - 2(3)^3 = -21 \text{ m}$

$\overline{v_x} = \dfrac{\Delta x}{\Delta t} = \dfrac{-21m - (7m)}{3s - 1s} = -14 \text{ m/s}$

note: $\overline{v_x} \neq \dfrac{v(1) + v(3)}{2} = \dfrac{0 + (-36 m/s)}{2} = -18 \text{ m/s}$

because a_x is not constant!

● ADDITIONAL PROBLEMS IN KINEMATICS IN ONE AND TWO DIMENSIONS CAN BE FOUND UNDER
    Constant Acceleration
    Projectile Motion
    Circular Motion

3
EQUILIBRIUM

NEWTON'S FIRST LAW

== 3-1

A cord is passed around a massless pulley and connected to the two masses. The mass m_2 is pulled by a constant horizontal force F as shown. If each of the block has to move with constant speed, what coefficient of kinetic friction is required between surfaces ? Assume that the same coefficient of kinetic friction between all surfaces, and that F = 5 nt, $m_2 = 2m_1$ =2kg.

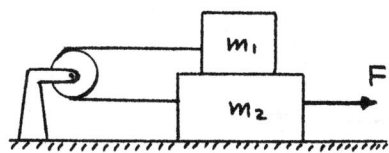

****

Moving with a constant speed implies that,

$$-T + \mu m_1 g = 0 \quad \text{for } m_1,$$

and

$$F - T - \mu(m_1+m_2)g - \mu m_1 g = 0 \quad \text{for } m_2,$$

where μ : the coefficient of kinetic friction

  T : tension in the cord. ($T_1 = T_2 = T$ in this problem)

$$\xleftarrow{T_1} \boxed{m_1} \xrightarrow{\mu m_1 g} \qquad \xleftarrow{\mu m_1 g} \overset{\xleftarrow{T_2}}{\boxed{m_2}} \xrightarrow{F}$$
$$\mu(m_1+m_2)g$$

Eliminating T from the two equations,

$$\mu = \frac{F}{(3m_1+m_2)g} = \frac{1}{9.8} \approx 0.102 \quad \leftarrow$$

EQUILIBRIUM OF A PARTICLE

3-2

A 75 lb. traffic light is suspended from two cables, which hang at 70° to the vertical, as shown:

Find the tension in the cables.

***

Draw the force diagram:

The upward-components of the two cables work together to counterbalance the force of gravity:

$$2T \sin 20° = 75 \text{ lb.}$$

$$T = \frac{75 \text{ lb}}{2 \sin 20°}$$

$$\boxed{T = 110 \text{ lbs}}$$

3-3

Knowing that the force applied to the three wires by the spring is given by the equation:
$$F_S = k\Delta x$$
where $\Delta x = 6.54$ cm is the displacement of the spring from its equilibrium position, and that the tension in the middle wire is $B = 18$ N; find the tensions A and C assuming the system to be in equilibrium.

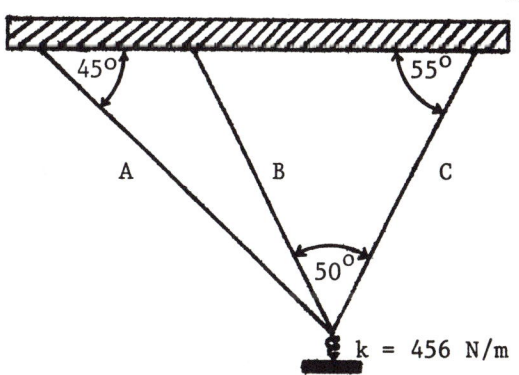

****

$\Sigma F_x = A\cos 135° + B\cos 105° + C\cos 55° + F_S \cos 270° = 0$
$= -.707A - .259B + .574C = 0$
(1) $= -.707A - 4.662N + .574C = 0$

$\Sigma F_y = A\sin 135° + B\sin 105° + C\sin 55° + F_S \sin 270° = 0$
$= .707A + .966B + .819C - F_S = 0 \; ; F_S = k\Delta x$
$= (456 \frac{N}{m})(.0654 m)$
$= 29.822 N$
$= .707A + (.966)(18N) + (.819)C - 29.822N = 0$
$= .707A + 17.388N + (.819)C - 29.822N = 0$
(2) $= .707A + (.819)C - 12.434N = 0$

Solve equation (1) for A: $A = .812C - 6.594N$
Substitution of this result into expression (2) gives us,
$(.707)(.812C - 6.594N) + (.819)C = 12.434N$
or
$C = (12.434N + 4.662N)/(.574 + .819) = 12.273N$

At this point, we are able to find A by substituting the value of C into either expression (1) or (2). The choice is completely arbitrary. Substitution of $C = 12.273N$ into expression (1) yields,

$-.707A - 4.662N + (.574)(12.273N) = 0$
or
$A = [(4.662N - (.574)(12.273N)]/(-.707) = 3.370 N$

3-4

Consider the weight "W" hanging from two ropes as shown below.
a) At what angle ϕ will the tensions T_1 and T_2 each be equal to the weight W.
b) Is this angle independent of the value of W?

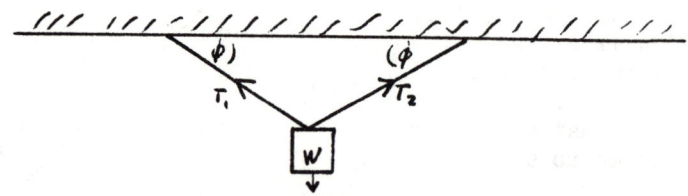

****

a) Sum the forces acting on 'W' along x:

$$\Sigma F_x = -T_1 \cos\phi + T_2 \cos\phi = 0 \qquad (1)$$

$$\therefore T_1 = T_2$$

Next, sum the forces acting on 'W' along y:

$$\Sigma F_y = T_1 \sin\phi + T_2 \sin\phi - W = 0 \qquad (2)$$

$$2 T_1 \sin\phi = W \qquad (3)$$

$$\sin\phi = \frac{W}{2T_1}$$

If $T_1 = W$, $\sin\phi = \frac{1}{2}$

$$\therefore \phi = 30°$$

b) Yes. For any 'W', $\phi = 30°$ will make the tension in the rope equal to W.

i.e. from (3) $T_1 = \frac{W}{2\sin\phi}$

and if $\phi = 30°$, $\sin\phi = \frac{1}{2}$

$$\therefore T_1 = W.$$

3-3

Knowing that the force applied to the three wires by the spring is given by the equation:
$$F_S = k\Delta x$$
where $\Delta x = 6.54$ cm is the displacement of the spring from its equilibrium position, and that the tension in the middle wire is $B = 18$ N; find the tensions A and C assuming the system to be in equilibrium.

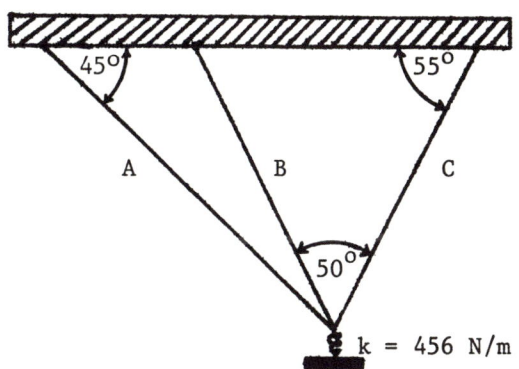

****

$\Sigma F_x = A\cos 135° + B\cos 105° + C\cos 55° + F_S \cos 270° = 0$
$\quad = -.707A - .259B + .574C = 0$
(1) $\quad = -.707A - 4.662N + .574C = 0$

$\Sigma F_y = A\sin 135° + B\sin 105° + C\sin 55° + F_S \sin 270° = 0$
$\quad = .707A + .966B + .819C - F_S = 0 \;;\; F_S = k\Delta x$
$\quad\quad\quad\quad\quad\quad\quad\quad\quad\quad\quad\quad\quad = (456 \tfrac{N}{m})(.0654 m)$
$\quad\quad\quad\quad\quad\quad\quad\quad\quad\quad\quad\quad\quad = 29.822 N$
$\quad = .707A + (.966)(18N) + (.819)C - 29.822N = 0$
$\quad = .707A + 17.388N + (.819)C - 29.822N = 0$
(2) $\quad = .707A + (.819)C - 12.434N = 0$

Solve equation (1) for A: $A = .812C - 6.594N$
Substitution of this result into expression (2) gives us,
$\quad (.707)(.812C - 6.594N) + (.819)C = 12.434N$
or
$\quad C = (12.434N + 4.662N)/(.574 + .819) = 12.273N$

At this point, we are able to find A by substituting the value of C into either expression (1) or (2). The choice is completely arbitrary. Substitution of $C = 12.273N$ into expression (1) yields,

$\quad -.707A - 4.662N + (.574)(12.273N) = 0$
or
$\quad A = [(4.662N - (.574)(12.273N)]/(-.707) = 3.370 N$

3-4

Consider the weight "W" hanging from two ropes as shown below.
a) At what angle ϕ will the tensions T_1 and T_2 each be equal to the weight W.
b) Is this angle independent of the value of W?

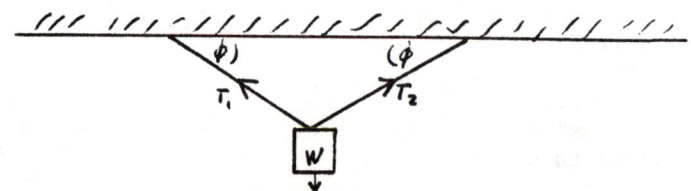

****

a) Sum the forces acting on 'W' along x:

$$\Sigma F_x = -T_1 \cos\phi + T_2 \cos\phi = 0 \qquad (1)$$

$$\therefore T_1 = T_2$$

Next, sum the forces acting on 'w' along y:

$$\Sigma F_y = T_1 \sin\phi + T_2 \sin\phi - W = 0 \qquad (2)$$

$$2 T_1 \sin\phi = W \qquad (3)$$

$$\sin\phi = \frac{W}{2 T_1}$$

If $T_1 = W$, $\sin\phi = \frac{1}{2}$

$$\therefore \phi = 30°$$

b) Yes. For any 'W', $\phi = 30°$ will make the tension in the rope equal to W.

i.e. from (3) $T_1 = \frac{W}{2 \sin\phi}$

and if $\phi = 30°$, $\sin\phi = \frac{1}{2}$

$$\therefore T_1 = W.$$

3-5

A knot is acted upon by two forces as shown. Find the force that will put the knot in equilibrium. Give magnitude and direction.

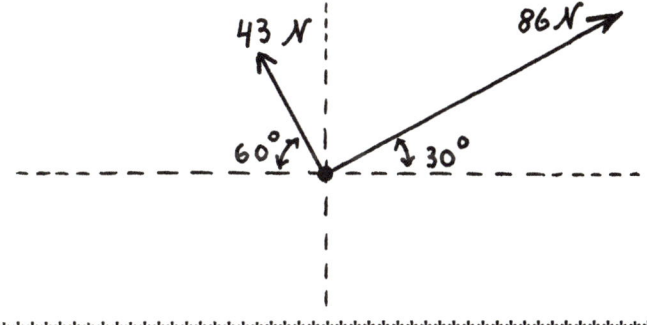

****

$\xrightarrow{+} \Sigma F_x = 0 = 86 \cos 30 - 43 \cos 60 + A_x$

$A_x = -52.5 = 52.5 \, N \leftarrow$

$+\uparrow \Sigma F_y = 0 = 86 \sin 30 + 43 \sin 60 + A_y$

$A_y = -80 = 80 \, N \downarrow$

$A = \sqrt{52.5^2 + 80^2} = 95.6$

$\tan \theta = \dfrac{52.5}{80} = 0.655 \qquad \theta = 33.2°$

$\underline{A = 95.6 \, N \; \measuredangle \, 33.2°}$

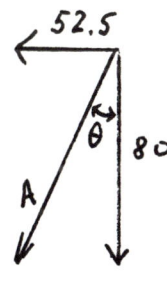

3-6

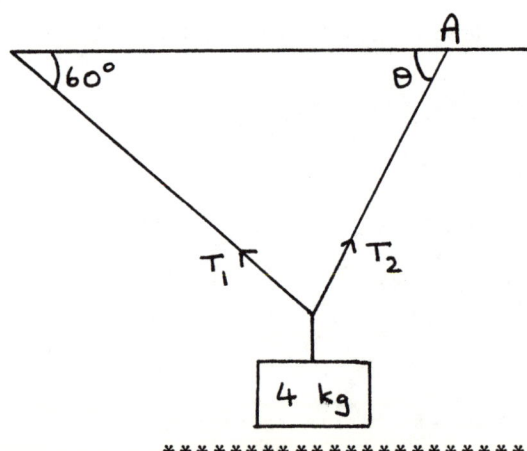

The end of the rope, A, can be moved horizontally, so the angle θ may change. Find the angle θ which makes T_2 exactly twice T_1 and calculate the tension T_1 in this configuration.

***

① $T_1 \sin 60° + T_2 \sin \theta = mg$ vertical dirctn.
② $T_1 \cos 60 = T_2 \cos \theta$ horiz. "
③ $T_2 = 2 T_1$

②: $T_1 \cos 60 = 2 T_1 \cos \theta$

∴ $\cos \theta = \dfrac{\cos 60}{2}$ ∴ $\theta = 75.52°$

∴ ①: $T_1 \left[\sin 60 + 2 \sin 75.52° \right] = mg$

∴ $T_1 = \dfrac{(4 \text{ kg})(9.8 \text{ m/s}^2)}{\left[\sin 60° + 2 \sin 75.52° \right]} = 13.99 \text{ N}$

3-7

A 10-kg mass is suspended from ceiling and wall by cords as shown. Find the tensions T_1 and T_2 in cords 1 and 2 respectively.

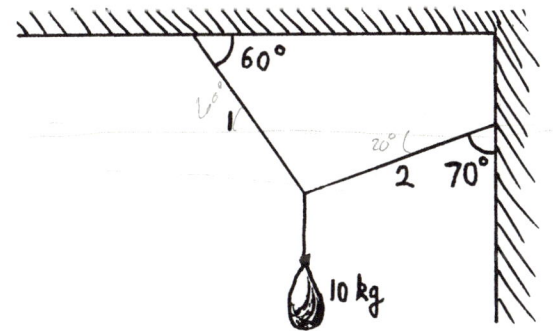

****

Free-body diagram:

$$\sum F_x = 0: \quad T_2 \cos 20° - T_1 \cos 60° = 0$$

$$T_2 = T_1 \frac{\cos 60°}{\cos 20°}$$

$$T_2 = 0.532\, T_1$$

$$\sum F_y = 0: \quad T_1 \sin 60° + T_2 \sin 20° - 98N = 0$$

$$T_1 \sin 60° + (0.532) T_1 \sin 20° = 98 N$$

$$T_1 = 94 \text{ N}$$
$$T_2 = 50 \text{ N}$$

FRICTIONAL FORCES

3-8

What force parallel to the incline is required to pull a 250 lb. weight up a 25° incline at constant speed if the coefficient of sliding friction is 0.20 and is constant?

****

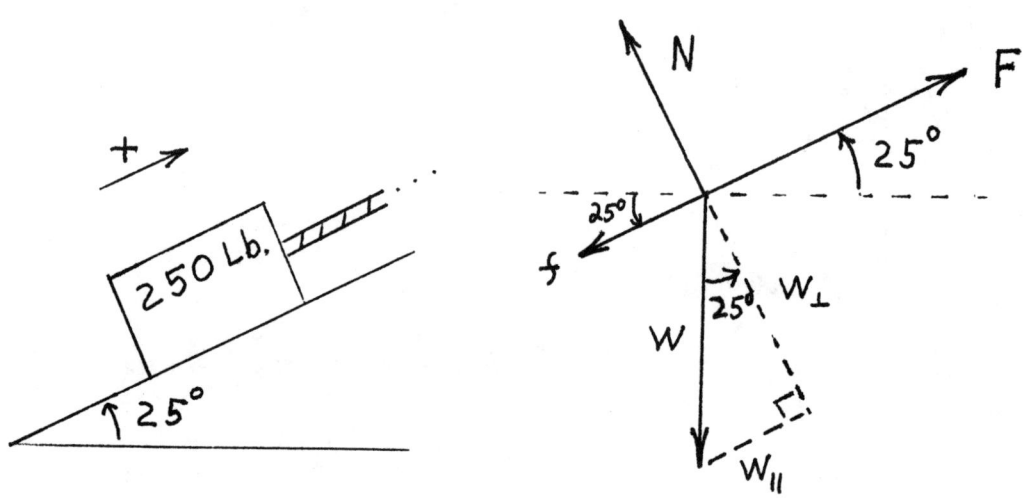

THE COMPONENTS OF $W = 250$ Lb, PARALLEL AND PERPENDICULAR TO THE INCLINE ARE:
$W_{\parallel} = 250$ Lb $SIN(25°) = 106$ Lb, AND $W_{\perp} = 250$ Lb $COS(25°) = 227$ Lb.

$\Sigma F_{\perp} = 0 \rightarrow N = W_{\perp} = 227$ Lb.
THEN FRICTION $= f = \mu N = 0.20 \times 227$ Lb $= 45$ Lb.

$\Sigma F_{\parallel} = 0 \rightarrow F - f - W_{\parallel} = 0$ OR $F = f + W_{\parallel} = 151$ Lb.

3-9

With the given values of the angles, the kinetic and static coefficients of friction seen in the drawing, what value of force, $\vec{F}$, in newtons, is required to pull the masses
$$m_1 = 9.5 \text{ kg}$$
$$m_2 = 11.5 \text{ kg}$$
at a constant velocity up the inclines ?

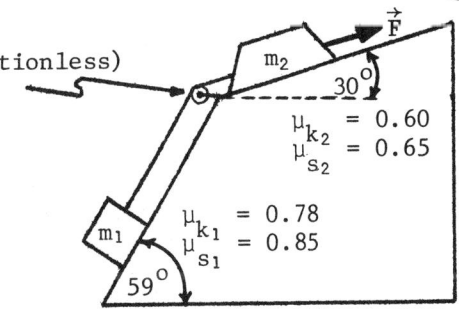

(frictionless)

$\mu_{k_2} = 0.60$
$\mu_{s_2} = 0.65$

$\mu_{k_1} = 0.78$
$\mu_{s_1} = 0.85$

****

Let us start by considering the forces on the two masses.

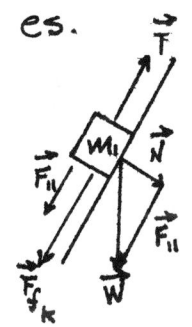

$\vec{F}_\parallel$ = parallel force to the incline's surface
   = $m_1 g \sin 59°$
   = $(9.5 \text{kg})(9.81 \text{m/sec}^2) \sin 59° = 79.884 N$

$\vec{N}$ = normal or contact force
   = $m_1 g \cos 59°$
   = $(9.5 \text{ kg})(9.81 \text{m/sec}^2) \cos 59° = 47.999 N$

$\vec{F}_{f_k}$ = kinetic frictional force = $\mu_{k_1} \vec{N}$
   = $(.78)(47.999 N) = 37.439 N$

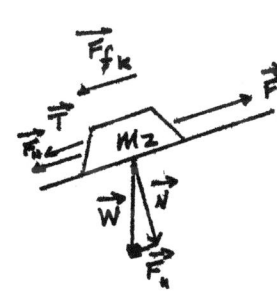

$\vec{F}_\parallel = m_2 g \sin 30° = (11.0 \text{kg})(9.81 \text{m/sec}^2) \sin 30°$
   = $53.955 N$

$\vec{N} = m_2 g \cos 30° = (11.0 \text{kg})(9.81 \text{m/sec}^2) \cos 30°$
   = $93.453 N$

$\vec{F}_{f_k} = \mu_{k_2} \vec{N} = (0.60)(93.453 N) = 56.072 N$

From the two previous force analyses, we are able to arrive at two equations, one for each mass:

m_1: $\vec{F}_\parallel + \vec{F}_{f_k} = \vec{T}$ or $\vec{T} = 79.884 N + 37.439 N = 117.323 N$

m_2: $\vec{F}_\parallel + \vec{F}_{f_k} + \vec{T} = \vec{F}$ or $\vec{F} = 53.955 N + 56.072 N + 117.323 N$
$$\vec{F} = 227.350 N$$

3-10

A small block of mass m rides atop a larger block that is moving horizontally. The coefficient of static friction between the blocks is μ_s.
  (a) If the magnitude of the acceleration of the lower block exceeds some limit, the small block will slide off the lower block. Derive an equation that gives this critical value of acceleration, assuming that the speed of the lower block is increasing.
  (b) Would your equation from part (a) also apply if the speed of the lower block is decreasing?

***

forces on upper block

$$\sum F_x = f_s = ma$$
$$\sum F_y = N - mg = 0 \Rightarrow N = mg$$

(a) Let $f_s = f_{S\,MAX} = \mu_s N = \mu_s mg \Rightarrow a = a_{MAX}$

Then $a_{MAX} = f_{S\,MAX}/m = \underline{\mu_s g}$

(b) YES --- $f_s = ma$ whether the speed is increasing or decreasing.

3-11

Two blocks move with constant velocity on a double incline, as shown. The cord connecting them is massless and inextensible, and is strung parallel to the surfaces of the inclines. The pulley over which the cord runs is massless and frictionless. The coefficient of kinetic friction between the blocks and the surface of the incline is 0.35, and block #1 moves down the plane (to the right). If block #2 weighs 50 N, what is the weight of block #1?

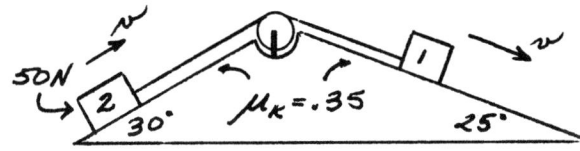

****

Step I: draw free-body diagrams for each block, isolating the block & identifying all the forces on it. Then set up Newton's first law. (Why first law? Because the block's aren't accelerating. How do I know? Because the problem says they move with CONSTANT velocity $\Rightarrow a=0 \Rightarrow \Sigma \vec{F}=0$).

Block 2:

$$\Sigma F_y = N_2 - W_2 \cos 30° = 0 \Rightarrow N_2 = W_2 \cos 30°= 43.3 \, N$$

$$\Sigma F_x = T - F_2 - W_2 \sin 30° = 0$$

$$\Rightarrow T = F_2 + W_2 \sin 30° = \mu_k N_2 + W_2 \sin 30°$$

$$= (0.35)(43.3 N) + (50 N)(0.5)$$

$$T = 40.2 \, N$$

(frictional force $= \mu_k N_2$)

Block 1

$$\Sigma F_y = N_1 - W_1 \cos 25° = 0 \Rightarrow N_1 = W_1 \cos 25°$$

$$\Sigma F_x = W_1 \sin 25° - F_1 - T = 0$$

$$\Rightarrow W_1 \sin 25° - \mu_k N_1 - T = 0$$

$$W_1 (\sin 25° - \mu_k \cos 25°) = T = 40.2 \, N$$

(frictional force $= \mu_k N_1$)

$$\Rightarrow W_1 = 40.2 / (.423 - (0.35)(.906))$$

$$\boxed{W_1 = 381 \, N}$$

THE MOMENT OF A FORCE · TORQUE

3-12

A non-uniform meter stick weighing 100 g supports a 400-g mass at the zero end and a 50-g mass at the other end. It is balanced on a fulcrum placed at the 20-cm mark. Find the center of gravity of this non-uniform meter stick.

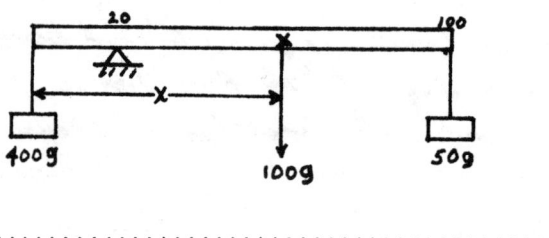

***

Since the 100-g weight of the meter stick can be considered to act at its center of gravity, we show it acting at a distance x from the zero end. Then moments around an axis through the fulcrum should balance:

$\sum M$ (clockwise): $100(x-20) + 50 \times 80 = 100x - 2000 + 4000 = 100x + 2000$

Counterclockwise: $400 \times 20 = 8000$

So, $100x + 2000 = 8000 \implies 100x = 6000$

or $x = 60$ cm

Determine the moment of force about point A:
(a) of the 30 N force;
(b) of the 40 N force;
(c) of the 55 N force.

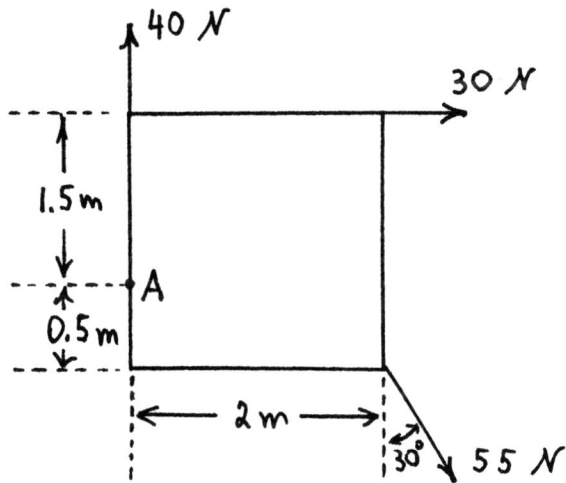

****

(a) $30(1.5) = \underline{\underline{45 \text{ Nm}}} \curvearrowright$

(b) $40(0) = 0$

(c) $55(\sin 30)(0.5) \curvearrowleft + 55(\cos 30)(2) \curvearrowright$

$\quad = 13.7 \curvearrowleft + 95.3 \curvearrowright = \underline{\underline{81.6 \text{ Nm}}} \curvearrowright$

CENTER OF GRAVITY

3-14

(a) Three particles have masses M_1 = 2 kg, M_2 = 3 kg, and M_3 = 5 kg. They lie along a line at positions described by coordinates x_1 = 3 m, x_2 = -4 m, and x_3 = 8 m. Find the position of the center of mass. (Show work.)

(b) Two particles have masses M_1 = 0.2 kg and M_2 = 0.3 kg. They have velocities V_1 = (2i + 3j) m/s and V_2 = (4i - j) m/s. Find the velocity of the center of mass of this two-particle system. (Show work.)

****

✓ (a)

$m_2 = 3\,kg$ at $-4(m)$, $m_1 = 2\,kg$ at $+3(m)$, $m_3 = 5\,kg$ at $+8(m)$ — x axis

$$X_{cm} = \frac{M_1 X_1 + M_2 X_2 + M_3 X_3}{M_1 + M_2 + M_3}$$

$$= \frac{6 + (-12) + 40}{2 + 3 + 5} = \frac{34\,kg \cdot m}{10\,kg} = \underline{3.4\,m}$$

(b) $\vec{V}_{cm} = \frac{M_1 \vec{U}_1 + M_2 \vec{U}_2}{M_1 + M_2} \quad \left[= (V_{cm})_x \vec{i} + (V_{cm})_y \vec{j} \right]$

$$= \frac{M_1 U_{1x}\vec{i} + M_1 U_{1y}\vec{j} + M_2 U_{2x}\vec{i} + M_2 U_{2y}\vec{j}}{M_1 + M_2}$$

$$= \underbrace{\left(\frac{M_1 U_{1x} + M_2 U_{2x}}{M_1 + M_2}\right)}_{(V_{cm})_x} \vec{i} + \underbrace{\left(\frac{M_1 U_{1y} + M_2 U_{2y}}{M_1 + M_2}\right)}_{(V_{cm})_y} \vec{j}$$

Substitution of numbers gives

$(V_{cm})_x = 3.2 \frac{m}{s}$ and $(V_{cm})_y = 0.60 \frac{m}{s}$

$$\therefore \vec{V}_{cm} = 3.2\, \vec{i} + 0.60\, \vec{j}$$

--3-15

Find the center of mass or gravity of a right triangle having shorter sides of length a and b.

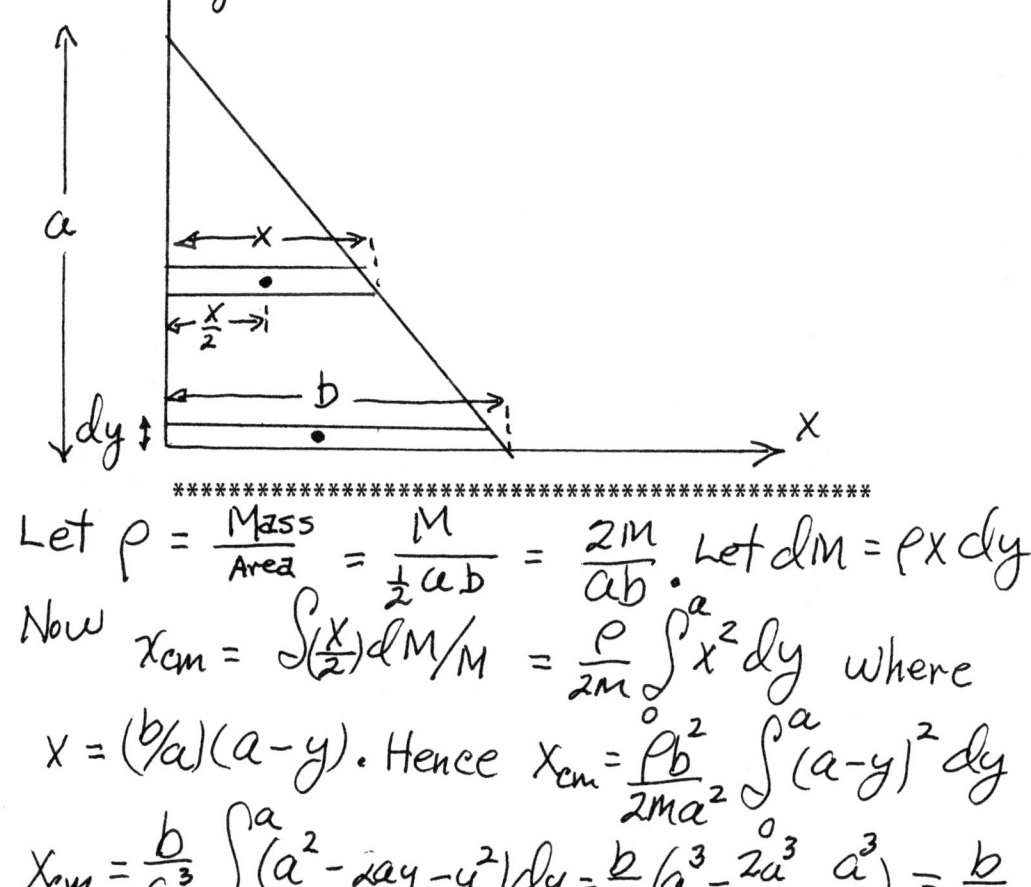

Let $\rho = \frac{Mass}{Area} = \frac{M}{\frac{1}{2}ab} = \frac{2M}{ab}$. Let $dM = \rho x\, dy$

Now $x_{cm} = \int (\frac{x}{2}) dM / M = \frac{\rho}{2M} \int_0^a x^2 dy$ where

$x = (b/a)(a-y)$. Hence $x_{cm} = \frac{\rho b^2}{2Ma^2} \int_0^a (a-y)^2 dy$

$x_{cm} = \frac{b}{a^3} \int_0^a (a^2 - 2ay - y^2) dy = \frac{b}{a^3}(a^3 - \frac{2a^3}{2} - \frac{a^3}{3}) = \frac{b}{3}$

Now consider y_{cm}

$$y_{cm} = \frac{\int y\,dM}{M} = \frac{\rho \int xy\,dy}{M} = \frac{2}{ab}\int_0^a y\left[\frac{b}{a}(a-y)\right]dy$$

$$y_{cm} = \frac{2}{a^2}\int_0^a y(a-y)\,dy = \frac{2}{a^2}\left[\frac{ay^2}{2} - \frac{y^3}{3}\right]_0^a$$

$$y_{cm} = \frac{2}{a^2}\left[\frac{a^3}{2} - \frac{a^3}{3}\right] = a - \frac{2a}{3} = \frac{a}{3}$$

EQUILIBRIUM OF RIGID BODIES

3-16

A RECTANGULAR DOOR IS HANGING BY THREE HINGES ALONG ITS LEFT SIDE, AT THE TOP, MIDDLE AND AT THE BOTTOM. THE DOOR IS A RECTANGLE, 2.0 METERS HIGH BY 1.0 METER WIDE. THE DOOR CENTER OF GRAVITY IS AT ITS GEOMETRICAL CENTER, AND ITS MASS IS 300 Kg. A 50 Kg BOY IS HANGING ON THE DOORKNOB, 10 Cm FROM THE RIGHT HAND EDGE. THE WEIGHT OF THE DOOR AND BOY IS DIVIDED EVENLY BETWEEN THE 3 HINGES. THE MIDDLE HINGE CANNOT SUPPORT A HORIZONTAL FORCE. FIND THE FORCES AT EACH HINGE.

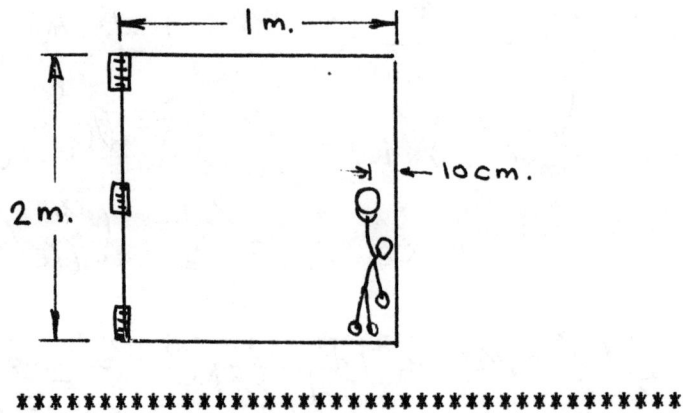

Force Diagram shown at right

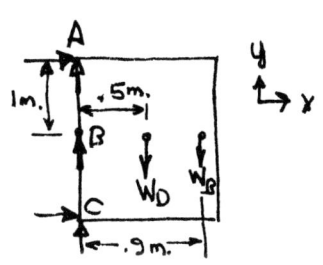

$W_D = 300 \text{ kg} \times 9.8 \text{ m/s}^2 = 2940 \text{ N}$

$W_B = 50 \text{ kg} \times 9.8 \text{ m/s}^2 = 490 \text{ N}$

Summing forces in x direction

$\Sigma F_x = 0 = A_x + C_x \qquad \therefore A_x = -C_x$

Summing forces in y direction

$\Sigma F_y = 0 = B_y + A_y + C_y - W_D - W_B$

$\therefore A_y + C_y + B_y = 2940 \text{ N} + 490 \text{ N}$

$\qquad\qquad\qquad\quad = 3430 \text{ N}$

But it is given that $A_y = C_y = B_y$

$\therefore 3 A_y = 3430 \text{ N}$

$\qquad A_y = 1143 \text{ N} = B_y = C_y$

Summing moments about hinge C

$\vec{\Sigma} T_C = .9\text{m} \times W_B + .5\text{m} \times W_D + 2\text{m} \cdot A_x = 0$

$\therefore A_x = \dfrac{-.9\text{m} \times 490 \text{ N} - .5\text{m} \times 2940 \text{ N}}{2\text{m}}$

$\qquad\quad = -956 \text{ N}$

$\therefore C_x = -A_x = 956 \text{ N}$

48 / Equilibrium

3-17

One end of a 100 N strut is against a vertical wall. Attached to the free end of the strut is a weight of W=50 N. Midway between the center of mass of the strut and the free end of the strut a support cable is attached at an angle θ=45°. What must be the minimum coefficient of friction between the wall and the strut for the system to remain in equilibrium?

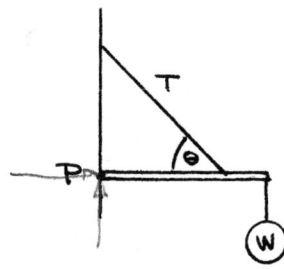

****

Describing problem:

Let S be weight of strut = 100N
     T the tension in the cable
     F the force by strut on wall at point P
     W = 50N

Setting up eqs.

From the 1st condition of equilibrium

$$F_x = T_x = T\cos\theta$$

$$F_y = W + S - T_y = W + S - T\sin\theta$$

Using the 2nd condition of equilibrium with pivot at P. Choose length of strut to be 4ℓ. Then CM at 2ℓ and cable attached at 3ℓ.

$$T_y(3\ell) = W(4\ell) + S(2\ell)$$

$$\Rightarrow T_y = \frac{4W + 2S}{3} \quad \Rightarrow T = \frac{4W + 2S}{3\sin\theta}$$

Then

$$F_y = W+S - \frac{4W+2S}{3\sin\theta}\sin\theta = \frac{S-W}{3}$$

$$F_x = \frac{4W+2S}{3\sin\theta}\cos\theta = \frac{4W+2S}{3\tan\theta}$$

So using eq. for friction

$$f = \mu F_N \quad \text{where} \quad f = F_y \,;\, F_N = F_x$$

$$\Rightarrow \mu = \frac{F_y}{F_x} = \frac{S-W}{4W+2S}\tan\theta$$

Substituting numbers:

$$\mu = \frac{100-50}{4(50)+2(100)}\tan(45°) = \frac{50}{400} = .125$$

3-18

A crude bridge across a stream on a hiking trail is constructed from a 14m long log having a uniform cross section and density, with a mass of 600kg. The log rests on supports at each end. An obese scoutmaster of mass 130kg stands on the log. The force on the left end support is 3400N. a) Find the force on the right end support. b) Find the position of the scoutmaster.

****

a.) $\sum f_{VERT} = 0$

$0 = (130 + 600)9.8 - 3400 - F$

$F = 3750\,N$

b.) $\sum \gamma = 0$ Take moments about the left end.

$$0 = (600*9.8)7 + (130*9.8)X - 3750*14$$

$$X = \frac{52500 - 41160}{1274} = 8.9 \, m$$

3-19

A uniform steel beam is 20 ft. long and weighs 400 lb. It is pivoted about its upper end. It is held at an angle of 30° to the vertical by a cable attached to its lower end. If the cable makes an angle of 45° above the horizontal, what is the force on the cable?

****

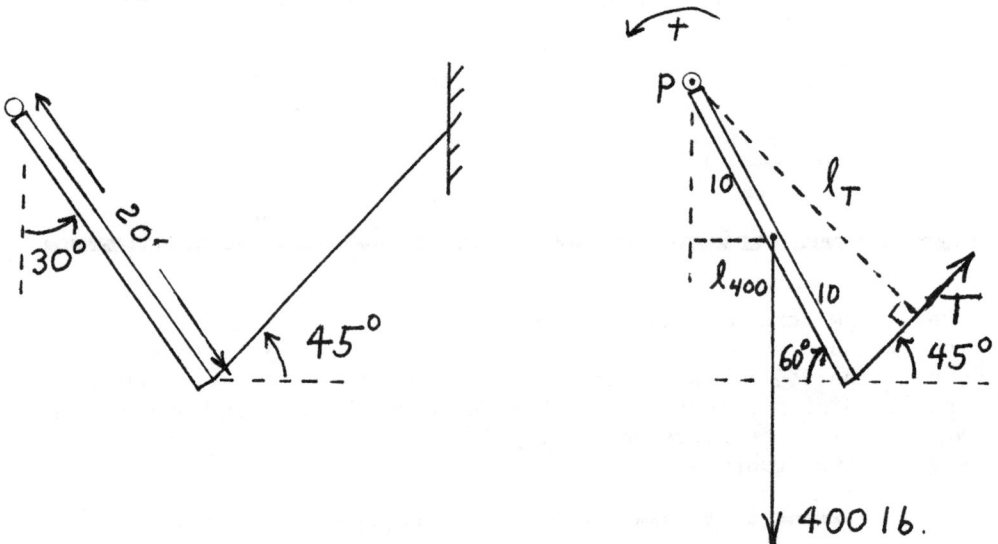

$\sum \tau_p = 0$ USING ℓ_T AND ℓ_{400} AS LEVER ARMS:

$$-400 \, \ell_{400} + T \ell_T = 0$$

$$-400 \, lb \times 10 ft \times SIN \, 30° + T \times 20 ft \times SIN(180-60-45) = 0$$

OR $19.3 \, T = 2000 \, lb$ SO $T = 104 \, lb.$

3-20

AB is a uniform board, 12 ft long and weighing 40 lb. End A is attached to the floor by a pivot. Force F, applied perpendicular to the board 4 ft from end B, supports the board at an angle of 30° with the floor.

(a) Find the size of force F.
(b) Find the horizontal and vertical components of the force exerted on the board by the pivot at A.

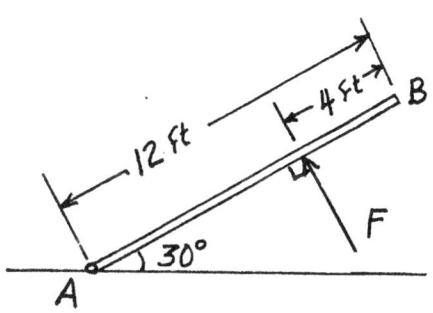

* *

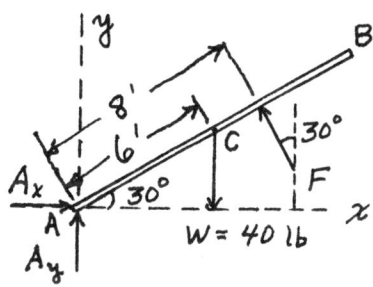

Draw a free-body diagram showing the 4 forces applied to board:
$\vec{F}$, ⊥ Board
$\vec{W}$ = 40 lb, down, applied at center
A_x, A_y, horizontal and vertical components exerted at A by pivot.

(a) Consider rotational equilibrium: Σ torques about A = 0.
A_x, A_y are through A and hence have zero torque
torque of W = W × moment arm = 40 × (6 cos 30°) = 208 lb-ft
torque of F = F × moment arm = F × 8 ft
The first is clockwise and the second is counterclockwise, so
$$8F - 208 = 0 \Rightarrow F = 26 \text{ lb}$$

(b) Consider translational equilibrium
$\Sigma F_x = 0$ $A_x - F \sin 30° = 0$
       $A_x = 26 \sin 30° = 13$ lb
$\Sigma F_y = 0$ $A_y + F \cos 30° = W$
       $A_y = 40 - 26 \cos 30° = 17.5$ lb.

Check: Calculate torque about center of board:
torque of W = 0
CW torque = torque of A_y = 17.5 × 6 cos 30° = 90.9 lb-ft
CCW = torque of A_x + torque of F
    = 13 × 6 sin 30 + 26 × 2 = 91 lb-ft = CW torque

3-21

A meter stick whose mass is uniformly distributed along its length is pivoted at its 5 cm mark by a horizontal frictionless peg. A mass (m = 100 grams) is hung at the stick's 90 cm mark. A light string is attached to the meter stick at its 15 cm mark to keep it horizontal, producing a tension of 14 N in the string which makes an angle of 130° with respect to the stick. (a) What is the mass of the meter stick? (b) What is the magnitude of the horizontal force exerted by the peg?

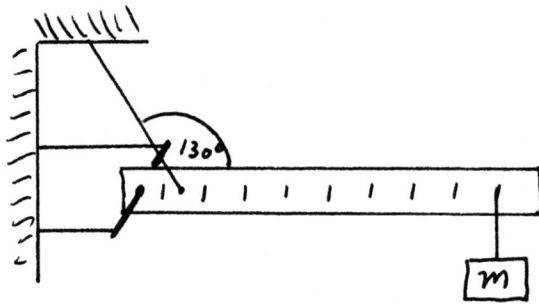

****

(a) Draw a free-body diagram for the stick

The force $M_s g$ belongs at the position of the center-of-mass for the stick and is therefore at $x = 50$ cm. Since we don't know P_1 and P_2 and we want to find M_s we'll take the torques (τ) about the pivot and get

$$\Sigma \tau_p = \Sigma \vec{r} \times \vec{F} = \Sigma r F \sin \phi = 0$$

$$(14)(0.1)\sin 130° - M_s(9.8)(0.45)\sin 90° - (.1)(9.8)(.85)$$
$$*\sin 90° = 0$$

$$M_s = \frac{(14)(.1)\sin 130° - (.1)(9.8)(.85)(1)}{(9.8)(.45)(1)} = \underline{\underline{0.054 \text{ kg}}}$$

(b) Using Newton's Second Law

$$\Sigma F_{horiz} = 0$$
$$P_2 - T \cos 50° = 0$$
$$P_2 = 14 \cos 50° = \underline{\underline{9.0 \text{ N}}}$$

3-22

A post 1.414m long, of weight 200N, is mounted on a hinge H so that it makes an angle of 45° with the horizontal. A weight of 500N is suspended at end B, and a cable AB supporting the system makes a 30° angle with the ground at point A. (a) What is the angle ABH? (b) Find the tension in the cable. (c) Find the magnitude of the force exerted by the hinge H on the post, and the angle this force makes with the horizontal. Assume a uniform post.

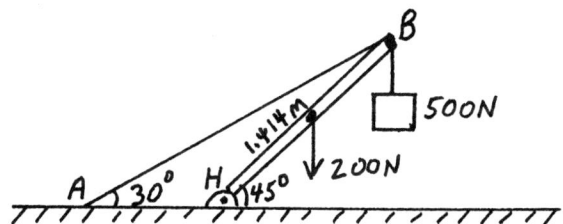

***

(a)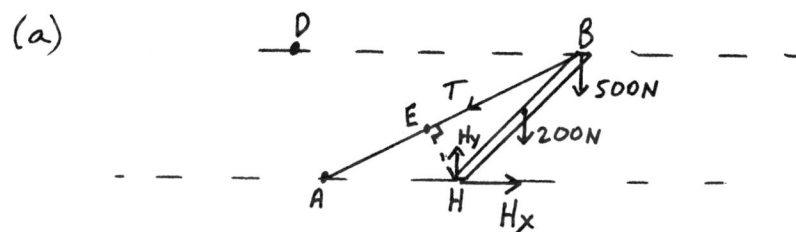

If D is an arbitrary point on a horizontal through B, then $\angle DBA = \angle BAH = 30°$. Also, $\angle DBH = 45°$. Therefore $\angle ABH = 45° - 30° = 15°$.

(b) Take axis for calculating torques at H. The moment arm of tension T is HE (drawn perpendicular to AB). $HE = 1.414 \sin 15°$. Then

$\Sigma \tau = 0 \quad T(1.414 \sin 15°) - 500(1.414 \cos 45°) - 200\left(\dfrac{1.414}{2} \cos 45°\right) = 0$

$T = 1639$ N

(c) $\Sigma F_x = 0 = H_x - T \cos 30° \qquad H_x = 1639 \cos 30° = 1420$ N
$\Sigma F_y = 0 = H_y - 200 - 500 - T \sin 30° \qquad H_y = 1520$ N

$H = \sqrt{H_x^2 + H_y^2} = 2080$ N $\qquad \tan \theta = \dfrac{H_y}{H_x} = \dfrac{1520}{1420} = 1.07$

$\theta = 46.9°$

3-23

A 20 kg uniform beam 12 meters long is pivoted at its end nearest a vertical wall on a hinge. It is held in a horizontal position by a cable attached to its center and to a point on the wall above the hinge such that the angle between the beam and the cable is 53°. (Use sin 53° = .800, cos 53° = .600). A 60 kg load is suspended from the beam at a point 9 meters from the hinge. The beam is in static equilibrium.
a) Find the tension in the cable, and
b) the horizontal and vertical components of the force on the hinge by the beam, in Newtons. Show your reasoning in getting these answers.

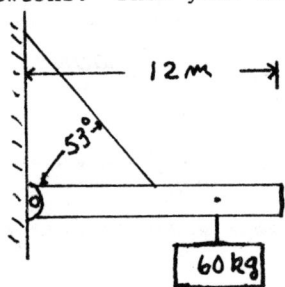

****

a) Free body diagram

$$\Sigma F_{vert} = 0: \quad V + T\sin 53° - 196 - 588 = 0$$

$$\Sigma F_{hor} = 0: \quad H - T\cos 53° = 0$$

$$\Sigma \Gamma_{pivot\ pt} = 0: \quad -196(6) - 588(9) + (T\sin 53°)6 = 0$$

$$\Rightarrow \quad T = \frac{(196)6 + 588(9)}{6 \sin 53°} = 1348\ N$$

b) $\quad H = 1348 \cos 53° = 809\ N$

$\quad V = 196 + 588 - 1348 \sin 53° = -294\ N$

The negative sign for V shows that V is **DOWN**

3-24

A uniform horizontal 10-ft bar weighs 100 lb. One end is attached to a vertical wall. A cable attached 1 ft from the far end of the bar is attached to the wall above the bar so that the cable makes a 60°-angle with the bar. A 500-lb load is attached to the far end of the bar.
a. Find the tension in the cable and the force the wall exerts on the bar.
b. Where should the cable be attached to the bar so that the will force is only horizontal? What will be the tension in the cable if the cable's point of attachment to the wall is not moved?

****

a.
$\Sigma F_x = 0 : F_x + T_x = 0;$ or $F_x = -T_x$
$\Sigma F_y = 0 : F_y + T_y = 500\,lb + 100\,lb = 600\,lb$
$\Sigma \Gamma = 0$ or $|\Sigma \Gamma_{cw}| = |\Sigma \Gamma_{ccw}|$:

$(100\,lb)(5\,ft) + (500\,lb)(10\,ft) = T_y (9\,ft)$

$T_y = \dfrac{500\,lb\,ft + 5000\,lb\,ft}{9\,ft} = 611\,lb$

$T_y = T \sin 60°; \quad T = \dfrac{T_y}{\sin 60°} = \dfrac{611\,lb}{0.866} = 706\,lb.$

$T_x = \dfrac{T_y}{\tan 60°} = 353\,lb; \quad \therefore F_x = -353\,lb,$ or $353\,lb$ out

$F_y = 600\,lb - T_y = 600\,lb - 611\,lb = -11\,lb$ or $11\,lb$ down

$F = \sqrt{F_x^2 + F_y^2} = \sqrt{(353\,lb)^2 + (-11\,lb)^2} = 353\,lb$

$\theta = \tan^{-1} \dfrac{-11\,lb}{353\,lb} = -1.8°$

$\therefore F = 353\,lb$ @ $1.8°$ below the bar.

b. If $F_y = 0$, $T_y = 500\,lb + 100\,lb = 600\,lb$.
$\therefore (100\,lb)(5\,ft) + (500\,lb)(10\,ft) = (600\,lb)(x)$

$x = \dfrac{5500\,lb\,ft}{600\,lb} = 9.17\,ft$ from the wall.

$y = (9\,ft) \tan 60° = 15.6\,ft$

New $\theta = \tan^{-1} \dfrac{15.6\,ft}{9.17\,ft} = 59.6°$

New $T = \dfrac{T_y}{\sin \theta} = \dfrac{600\,lb}{\sin 59.6°} = 696\,lb.$

3-25

A uniform ladder 20 ft long leans against a vertical wall in such a way that it makes an angle of 60° with the horizontal ground. The ladder weighs 50 lb. The coefficient of friction between the base of the ladder and the ground is 0.40.
a. How far up the ladder can a 180-lb painter go before the ladder begins to slip?
b. What coefficient of friction between the base of the ladder and the ground would be necessary for the painter to be able to go all the way up the ladder?

****

a. $G_v = Wt_p + Wt_\ell = 180\text{ lb} + 50\text{ lb} = 230\text{ lb}$
$G_h = f = \mu G_v = 0.4(230\text{ lb}) = 92\text{ lb}$

Take reference at wall:
$|\Sigma \Gamma_{cw}| = |\Sigma \Gamma_{ccw}|$
$(50\text{ lb})(5\text{ ft}) + (180\text{ lb})(x) + (92\text{ lb})(20\text{ ft})\sin 60° = (230\text{ lb})(10\text{ ft})$
$250\text{ lb·ft} + (180\text{ lb})x + 1593\text{ lb·ft} = 2300\text{ lb·ft}$

$x = \dfrac{2300\text{ lb·ft} - 1843\text{ lb·ft}}{180\text{ lb}} = 2.5\text{ ft from left}$

OR $10\text{ ft} - 2.5\text{ ft} = 7.5\text{ ft from right}$

$\ell \cos 60° = 7.5\text{ ft}; \ \ell = 15\text{ ft up ladder.}$

b. For painter at top of ladder, painter's lever arm = 0.

$\therefore |\Sigma \Gamma_{cw}| = |\Sigma \Gamma_{ccw}|$
$(50\text{ lb})(5\text{ ft}) + (180\text{ lb})(0) + \mu(230\text{ lb})(20\text{ ft})\sin 60° = (230\text{ lb})(10\text{ ft})$
$250\text{ lb·ft} + \mu(3984\text{ lb·ft}) = 2300\text{ lb·ft}$
$\mu = \dfrac{2300\text{ lb·ft} - 250\text{ lb·ft}}{3984\text{ lb·ft}} = 0.52.$

3-26

A 75 kg beam is mounted on a hinge and is supported by ropes as shown in the sketch. The mass of the suspended weight is 50 kg. Calculate the tensions in the ropes and the force on the hinge at the base of the beam.

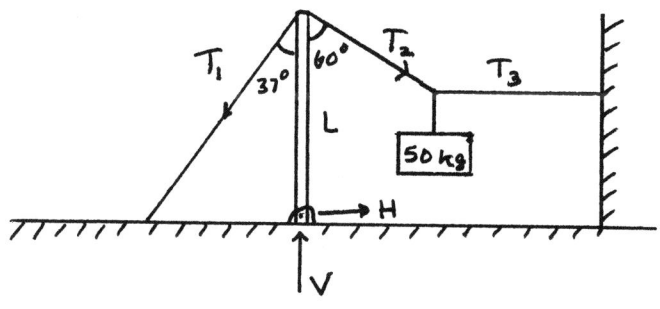

***

Let V, H be the forces on the bar hinge.

① $\sum_{bar} F_y = -T_1 \cos 37° - T_2 \cos 60° + V - 75(9.8)$ Nt. $= 0$

$\quad = -\frac{4}{5} T_1 - \frac{1}{2} T_2 + V - 73.5$ Nt $= 0$

② $\sum_{bar} F_x = H + T_2 \sin 60° - T_1 \sin 37° = 0 = H + \frac{\sqrt{3}}{2} T_2 - \frac{3}{5} T_1$

③ $\sum_{base} \tau = T_2 L \sin 60° - T_1 L \sin 37° = 0$

So $\frac{3}{5} T_1 = \frac{\sqrt{3}}{2} T_2$ or $T_1 = \frac{5\sqrt{3}}{6} T_2$

From ① $V = \frac{4}{5} T_1 + \frac{1}{2} T_2 + 73.5 = \frac{4}{5} \left(\frac{5\sqrt{3}}{6} T_2 \right) + \frac{1}{2} T_2 + 73.5$

$V = (1.15 + 0.5) T_2 + 73.5 = 1.65 T_2 + 73.5$ Nt.

$H = \frac{3}{5} T_1 - \frac{\sqrt{3}}{2} T_2 = \frac{3}{5} \frac{5\sqrt{3}}{6} T_2 - \frac{\sqrt{3}}{2} T_2 = \underline{0}$

Now apply equilibrium conditions to string junction

$\sum F_x = T_3 - T_2 \cos 30° = 0 \quad$ so $T_3 = \frac{\sqrt{3}}{2} T_2$

$\sum F_y = T_2 \sin 30° - 50(9.8)$ Nt $= 0 \quad$ so $\underline{T_2 = 980 \text{ Nt}}$

$T_3 = T_2 (.866) = 980(.866) = \underline{848 \text{ Nt}}$

$T_1 = \frac{5\sqrt{3}}{6} T_2 = 1.44(980) = \underline{1410 \text{ Nt}}$

$V = 1.65(980) + 73.5 = 1617 + 74 = \underline{1691 \text{ Nt}}$

3-27

A homogeneous meter stick is held by four fingers, one at the 0.0 cm location and three at the 100.0 cm location under the horizontal meter stick. A 1.0 kg mass placed on the meter stick makes it possible for the load to be equally divided among the four fingers if the mass of the meter stick is neglected. Where is the 1.0 kg mass located on the meter stick?

****

Let F = force exerted by any one finger.

Equal |torques| $\Rightarrow$ Rotational equilibrium

$$x F = (100 cm - x) 3F$$

$$x = 300 cm - 3x \quad \Rightarrow \quad 4x = 300 cm$$

$$x = \frac{300 cm}{4} \quad \Rightarrow \quad \boxed{x = 75 cm}$$

The 1.0 kg mass is located at the 75.0 cm position on the meter stick where it exerts a force of $Mg = (1.0 kg)(9.8 m/sec^2) = 9.8$ newtons downward on the meter stick.

3-28

A boom 6 m long weighing 200 N supports a 1200 N load with the aid of a cable as shown below. Find the tension T and the force exerted by the hinge.

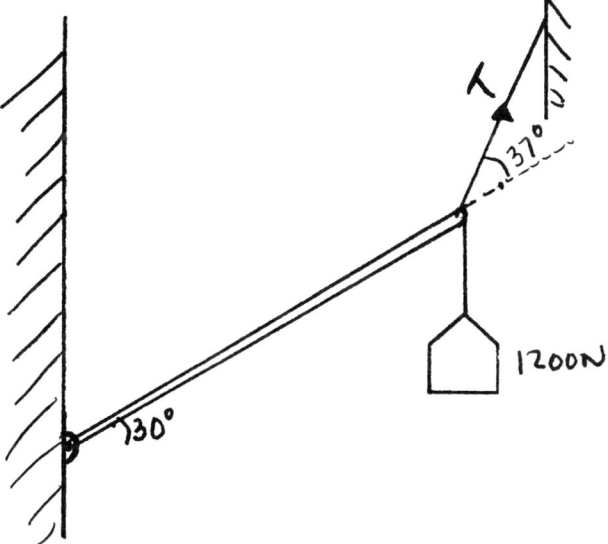

****

$\Sigma F_x = F_H + T\cos 67° = 0$ $\quad F_H$ = horizontal component of hinge

$\Sigma F_y = F_V + T\sin 67 - 1200 N - 200 N = 0$

$\quad F_V$ = vertical component of hinge

$\Sigma \tau_{hinge} = 200N(3m)\cos 30° + 1200N(6m)\cos 30° - T(6m)\sin 37°= 0$

$T(6m)\sin 37 = 200N(3m)\cos 30° + 1200N(6m)\cos 30°$

$T = \dfrac{519.6 \text{ N-m} + 6235.4 \text{ N-m}}{(6m)(\sin 37°)} = 1870.7 \text{ N}$

$F_V = -(1870.7N)\sin 67 + 1400N = -322 N$

$F_H = -(1870.7N)\cos 67 = -731 N$

$F_{hinge} = \sqrt{(322N)^2 + (731N)^2} = 799N$ at $204°$ counter-clockwise from the +x-axis.

3-29

A 100-lb, 16-ft uniform ladder rests against a smooth wall and makes an angle of 60° with the floor. The floor is also perfectly smooth. Can a 180-lb man stand safely on the ladder, 6 feet from the top without causing slipping if his son exerts a horizontal force of 120 lbs at a point 4 feet from the midpoint as shown?

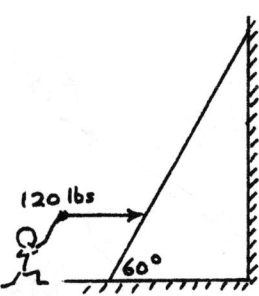

***

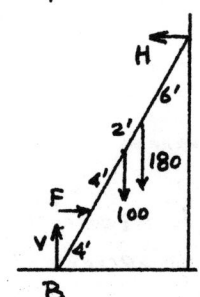

Since the wall and the floor are both smooth, there is no force component along the surfaces at A and B. Suppose that the boy exerts a horizontal force of F lbs. Then $\Sigma(\text{Torque})_A = 0$ gives,

$(3)(180) + (4)(100) + (6\sqrt{3})F - 8V = 0$,

where $V = 180 + 100$. ($F = H$, not needed here)

$\therefore F \cong 125.1$ lbs > 120 lbs.

Therefore, the answer is no.

3-30

A uniform beam 5.0 m long and weighing 500. N is attached at one end to a horizontal floor by a frictionless hinge. The other end of the beam is attached by a horizontal rope to an adjacent wall in such a fashion that the beam makes an angle of 53.13° with the horizontal. Three meters from the hinge a freely hanging weight, W, is attached to the beam as shown in the figure below.
  (a) How heavy a load, W, can be supported if the rope has a breaking strength of 2000. N?
  (b) What horizontal and vertical force must the hinge supply?

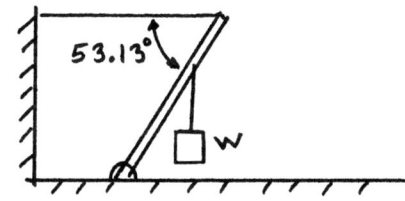

***

Let B be the weight of the beam, T the rope tension, and F_x and F_y the horizontal and vertical hinge forces respectively.
(Note: B acts at 2.5 m from hinge, given angle is for 3:4:5 right triangle.)

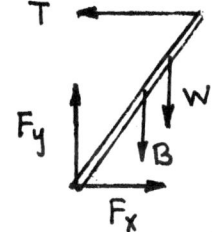

Free Body Diagram

In equilibrium the sum of force and torque components are independently zero.

(1) $\sum F_x = F_x - T = 0$

(2) $\sum F_y = F_y - B - W = 0$

Torques about hinge (3) $\sum \tau = 1.5 B + 1.8 W - 4 T = 0$

From (3) $W = \dfrac{4T - 1.5B}{1.8} = \dfrac{8000 \text{ Nm} - 750 \text{ Nm}}{1.8 \text{ m}} = \underline{\underline{4030 \text{ N}}}$

From (1) $F_x = T = \underline{\underline{2000 \text{ N}}}$

From (2) $F_y = B + W = \underline{\underline{4530 \text{ N}}}$

FRICTIONAL FORCES

3-31

Each block has the same weight, 10 N. The coefficient of friction is 0.25 for all surfaces. The bodies A and C are sliding down the plane. Body B is held stationary by a rope. Draw a free body diagram of each block. <u>Do not solve</u>.

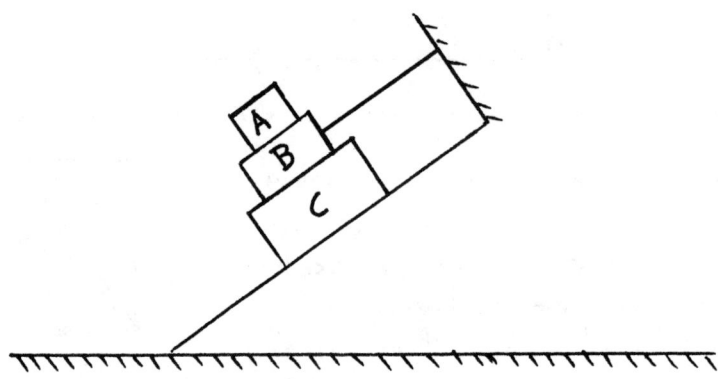

4
PLANE MOTION

NEWTON'S SECOND LAW

━━━ 4-1

I stand at rest where free-fall acceleration is 9.8 m/s² and I hold a 0.3 kg ball in my hand.

(a) If I move this ball vertically upward with a <u>constant velocity</u> of 4 m/s, what force (vector) must I exert on it?
(b) If I move this ball horizontally with a <u>constant acceleration</u> of 4 m/s², what force (vector) must I exert on it?
(c) If I move this ball at a <u>constant acceleration</u> of 4 m/s² in a direction 30° up from the <u>horizontal</u>, what force (vector) must I exert on it?

* *

Let x-axis (unit vector $\hat{\imath}$) be horizontal
    y-axis (unit vector $\hat{\jmath}$) be vertically up.
On the ball there are <u>two</u> forces:
  (i) the gravity force of Earth, Weight = $mg = 0.3 \cdot 9.8$
      $= 2.94$ N, down: $\vec{W} = -2.94\,\hat{\jmath}$ N.
  (ii) Force $\vec{F}$, exerted by me, to be found.
According to Newton:
  Vector sum of forces on ball
                    = (mass of ball) × (acceleration of ball)
or, here $\vec{F} + \vec{W} = m\vec{a} \Rightarrow \vec{F} - 2.94\,\hat{\jmath} = 0.3\,\vec{a}$
    so $\vec{F} = 0.3\,\vec{a} + 2.94\,\hat{\jmath}$, N

(a) $\vec{v} = 4\hat{j}$ m/s = const. $\therefore \vec{a} = 0$ and
$\vec{F} = 0.3 \cdot (0) + 2.94\hat{j} = 2.94\hat{j}$, N

When any object moves with any constant velocity ($\vec{v}$), zero <u>resultant</u> force is required, regardless of mass, speed, or direction. Since there are only two forces on the ball, they must be equal in size, opposite in direction.

(b) $a_x = 4$ m/s², $a_y = 0$, so $\vec{a} = 4\hat{i}$ m/s²
$\therefore \vec{F} = 0.3 \cdot 4\hat{i} + 2.94\hat{j} = 1.2\hat{i} + 2.94\hat{j}$, m/s²
I must exert a force $\sqrt{(1.2)^2 + (2.94)^2} = 3.18$ N
at angle $\theta = \text{Tan}^{-1}(1.2/2.94) = 22.2°$ with the vertical.

$\vec{F} + \vec{W} = m\vec{a}$
$W = 2.94$
$m\vec{a} = 1.2$

May also be solved by drawing a triangle showing that $m\vec{a}$ (known) is the vector sum of $\vec{W}$ (known) and $\vec{F}$ (unknown). The two legs of a right triangle are known, and the hypotenuse F and angle θ can be found by simple trigonometry.

(c)
$a_x = 4 \cos 30° = 3.46$; $a_y = 4 \sin 30° = 2.00$
so $\vec{a} = 3.46\hat{i} + 2.00\hat{j}$, m/s²
$\vec{F} = 0.3(3.46\hat{i} + 2.00\hat{j}) + 2.94\hat{j}$
$= 1.04\hat{i} + 3.54\hat{j}$, N.

Hence a force of $\sqrt{(1.04)^2 + (3.54)^2} = 3.96$ N at an angle θ, with the vertical, of $\text{Tan}^{-1}(1.04/3.54) = 16.4°$

$\vec{F} + \vec{W} = m\vec{a}$
$W = 2.94$
$ma = 1.2$

Again, the vector triangle showing $m\vec{a} = \vec{F} + \vec{W}$ can be constructed, as shown. The two legs are again known, but the angle between them is now 120° (= 90° + 30°). The scalene triangle can be solved trigonometrically (the law of cosines determines F, and then the law of sines gives θ) — but who wants to do that? The solution by components avoids <u>ever</u> having to solve scalene triangles!

Newton's Second Law / 65

4-2

Forces $\vec{F}_1$ and $\vec{F}_2$ act simultaneously on a particle of mass m = 1.5 (Kg). $\vec{F}_1$ has magnitude 12 Newton (N) and is directed 30° below the horizontal, while $\vec{F}_2$ is 5 (N) and directed 45° above the horizontal.

(a) Find the magnitude and direction of the resultant force.
(b) Find the magnitude and direction of the acceleration $\vec{A}$ of the particle.

****

(a) $\vec{F} = F_x \vec{i} + F_y \vec{j}$

Here, F_x and F_y are the components of the resultant force $\vec{F}$.

$F_x = F_{1x} + F_{2x} = F_1 \cos 30° + F_2 \cos 45°$
$\qquad = 12 \times 0.866 + 5 \times 0.707 = 13.9$ N

$F_y = F_{1y} + F_{2y} = -F_1 \sin 30° + F_2 \sin 45°$
$\qquad = -12 \times 0.5 + 5 \times 0.707$

$F_y = -2.5$ N

$\therefore F = \sqrt{F_x^2 + F_y^2} = \sqrt{(13.9)^2 + (-2.47)^2} = 14$ N

$\tan \theta = \dfrac{F_y}{F_x} = -\dfrac{2.5}{14} = -0.18$

$\theta = \tan^{-1}(-0.18) = -10°$, i.e., 10° below the horizontal.

(b) $\vec{a} = \dfrac{\vec{F}}{m}$

$\therefore \vec{a}$ is in same direction as $\vec{F}$

Also, magnitude $a = \dfrac{F}{m} = \dfrac{14 N}{1.5 kg} = 9.3$ m/s²

4-3

A student wishes to determine the coefficient of static friction between a block of material and a wooden board. He places the board on a table, places the block on the board, and gradually raises one end of the board. When the end of the board has been raised 20 cm the block slides 77.3 cm down the entire length of the board in 1.6 seconds. Find:

(a) the coefficient of static friction;
(b) the coefficient of kinetic friction;
(c) the angle for which the velocity of the block will be constant.
(d) How hard must one press on the block, perpendicular to the inclined surface to prevent it from sliding down a 30° incline?

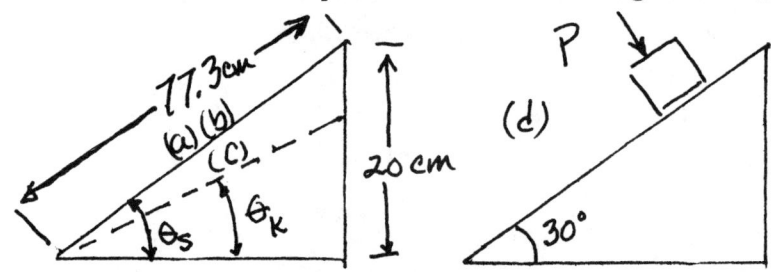

****

a) For the static situation ($v=0$) Newton's 2nd Law gives

$$mg \sin\theta_s = \mu_s mg \cos\theta_s$$

$$\mu_s = \tan\theta_s = \tan\left[\sin^{-1}\frac{20}{77.3}\right]$$

$$= \tan(15°) = 0.27$$

b) Once the block begins to slide, it will accelerate down the board and Newton's 2nd Law says

$$mg \sin\theta_s - \mu_K mg \cos\theta_s = ma$$

* $\sin(15°) - \mu_K \cos(15°) = \dfrac{a}{g}$

But we can find a since

$$77.3^m = \tfrac{1}{2} a (1.6s)^2$$

$$a = 60 \text{ cm/s}^2$$

So ∗ becomes

$$.27 - \mu_K(.97) = \frac{60 \text{ cm/s}^2}{980 \text{ cm/s}^2} = 0.06$$

$$\mu_K = 0.22$$

c) If $v = $ constant

$$mg \sin\theta_K = \mu_K \, mg \cos\theta_K$$

$$\mu_K = \tan\theta_K = 0.22$$

Therefore $\theta_K = 12.5°$

d) Without the additional normal force P we find

Normal $= mg \cos(30°) = .87 mg$

and friction force $= \mu_s N = (.27)(.87) mg = .23 mg$

Since the component of weight acting down the incline is

$$mg \sin(30°) = 0.5 mg$$

we must supply an additional friction force of $0.27 mg$ to keep the block stationary. This means we must increase the normal force by an amount

$$P = \frac{\Delta(\text{friction})}{\mu_s} = \frac{0.27 mg}{0.27} = mg$$

4-4

A 50 kg block is held at rest on a surface which is inclined at an angle of 60° with respect to the horizontal and has a coefficient of kinetic friction of 0.3 and a coefficient of static friction of 0.4.

Find: (a) the magnitude and direction of the acceleration when the block is released;

(b) the acceleration if the block had originally been traveling up the incline.

(c) Suppose that a force, F = 100 Newtons, is now exerted on the block parallel to the incline, directed up the incline. Find the acceleration of the block.

(d) Suppose that the force in part (c) had been directed along the horizontal, rather than parallel to the incline, so as to push the block up the incline. What is the acceleration?

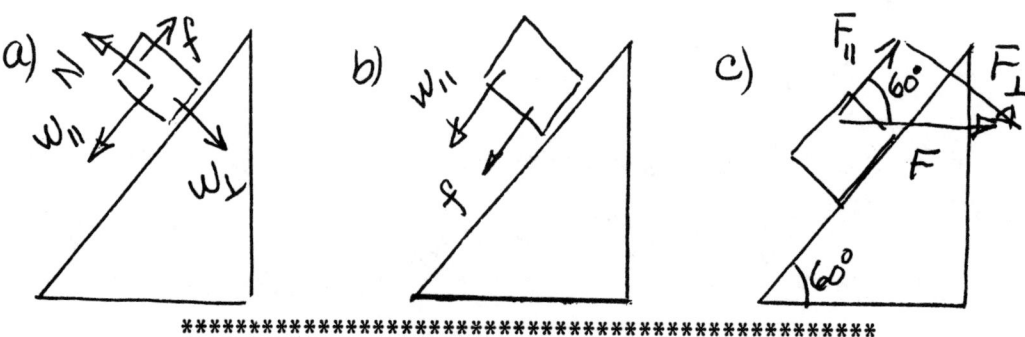

***

a) $W_\parallel = mg \sin 60° = (.87)(500 \text{ New}) = 435 \text{ New}.$
$N = W_\perp = mg \cos 60° = (.50)(500) = 250 \text{ New}.$

Find first the static friction force

$f_s = \mu_s N = (0.4)(250) = 100$ clearly not enough to keep block from sliding down the incline therefore

$mg \sin 60° - \mu_k mg \cos 60° = ma$
$435 - 75 = 360 = 50a$
$a = 7.2 \text{ m/s}^2 \text{ (down)}$

b) ~~If~~ the block had been initially travelling up the incline

$mg \sin 60° + \mu_k mg \cos 60° = ma$

$$a = (435+75)/50 = 10.2 \, m/s^2 \, (\text{down})$$

c) If block was initially moving up, for example

$$100 - 435 - 75 = 50a$$
$$a = -8.2 \, m/s^2 \, (\text{down})$$

But if block was initially moving down

$$a = -5.2 \, m/s^2 \, (\text{down})$$

d) If block is initially moving up incline

$$F_{\parallel} - mg\sin 60° - \mu_k N = ma$$

where $N = mg\cos 60 + F\sin 60° = 250 + 87 = 337$
$F_{\parallel} = F\cos 60° = 50$

and therefore

$$50 - 435 - (0.3)(337) = 50a$$
$$a = -9.7 \, m/s^2 \, (\text{down})$$

But if block had initially been moving down

$$a = -5.7 \, m/s^2 \, (\text{down})$$

In the above, note that a downward acceleration can be either positive or negative depending on how you chose to write these equations.

4-5

A mass $m_1 = 2.2$ kg moves on a frictionless inclined plane at an angle of 30° above the horizontal, as shown in the diagram. It is connected by a light string that passes over a frictionless pulley of negligible mass to another mass $m_2 = 2.7$ kg which hangs vertically without touching anything.
a) Draw separate free body diagrams for each of the masses.
b) Compute the magnitude and direction of the acceleration of m_1.
c) Calculate the tension in the string.

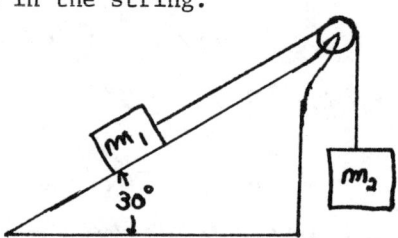

****

a) Free body diagrams:

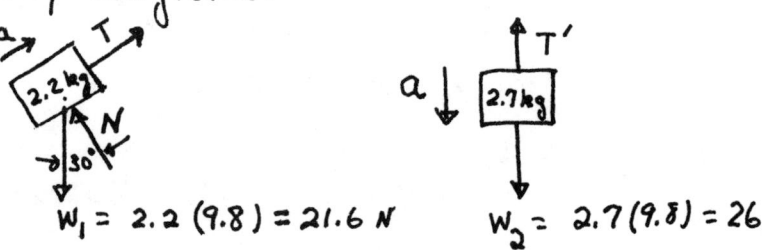

$W_1 = 2.2 (9.8) = 21.6$ N $W_2 = 2.7 (9.8) = 26.5$ N

b) neglecting mass of string $T = T'$

Newton's 2nd Law: Mass 1   use components ∥ and ⊥ to plane

$\sum F_\perp = 0$: $N - 21.6 \cos 30° = 0$

$\sum F_\parallel = m_1 a_\parallel$: $T - 21.6 \sin 30° = 2.2 a$

Newton's 2nd Law Mass 2   assume down is positive

$\sum F_{vert} = m_2 a_{vert}$: $26.5 - T = 2.7 a$

Eliminate T: $-21.6 \sin 30° + 26.5 = 4.9 a$

$a = \dfrac{15.7}{4.9} = 3.20$ m/s²

c) $T = 26.5 - 2.7 (3.2) = 17.8$ N

4-6

(a) Calculate the magnitude and direction of the acceleration of the 10 kg block.
(b) Calculate the tension in the cord between the 3 kg and the 2 kg blocks.
Use the 'Free Body Diagram Method'.
All the surfaces are smooth.

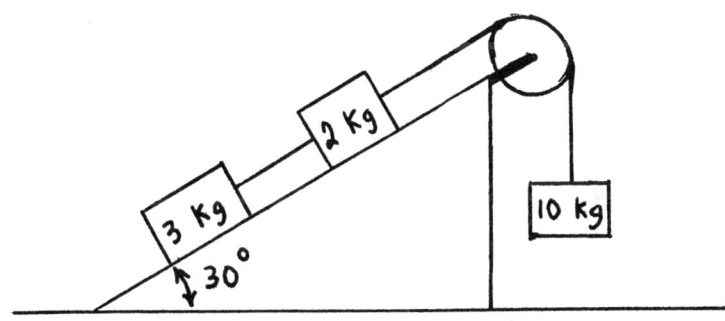

****

(a)

$+\nearrow \Sigma F_x = ma$

$T - 3(9.8)\sin 30 - 2(9.8)\sin 30 = (2+3)a$

$+\downarrow \Sigma F_y = ma$

$10(9.8) - T = 10a$

Adding above two equations the tension T drops out and we obtain: $10(9.8) - 5(9.8)\sin 30 = 15a$ and $\underline{\underline{a = 4.9 \frac{m}{s^2} \downarrow}}$

(b)

$+\nearrow \Sigma F_x = ma$

$T_1 - 3(9.8)\sin 30 = 3(4.9)$

$\underline{\underline{T_1 = 29.4 N \nearrow 30°}}$

4-7

A 640-lb crate rests on a hill whose angle of incline is 30°. The coefficient of static friction equals that of kinetic friction equals 0.40.
a. Find the object's acceleration.
Suppose a rope is attached to the crate and a 50-lb force is applied up the hill and parallel with the hill.
b. Find the object's acceleration.
c. If the force is increased to 200 lb, find the object's acceleration.
d. If the force is increased to 600 lb, find the object's acceleration.

****

a.
$Wt_\perp = Wt \cos 30° = (640\,lb)(0.866) = 554\,lb$
$Wt_{||} = Wt \sin 30° = (640\,lb)(0.50) = 320\,lb$
$m = \dfrac{Wt}{g} = \dfrac{640\,lb}{32\,ft/s^2} = 20\,slug$
$\Delta F = Wt_{||} - f = Wt_{||} - \mu Wt_\perp = 320\,lb - (0.40)(554\,lb) = 98\,lb$
$a = \dfrac{\Delta F}{m} = \dfrac{98\,lb}{20\,slug} = 4.9\,ft/s^2$ downhill.

b.
$\Delta F = Wt_{||} - f - F$
$= 320\,lb - 222\,lb - 50\,lb = 48\,lb$
$a = \dfrac{\Delta F}{m} = \dfrac{48\,lb}{20\,slug} = 2.4\,ft/s^2$ downhill.

c. $\Delta F(downhill) = Wt_{||} - f - F = 320\,lb - 222\,lb - 200\,lb < 0$
∴ no ΔF downhill

$\Delta F(uphill) = F - Wt_{||} - f = 200\,lb - 320\,lb - 222\,lb < 0$
∴ no ΔF uphill
∴ $a = 0$; or crate will not move.

d. $\Delta F = F - Wt_{||} - f = 600\,lb - 320\,lb - 222\,lb = 80\,lb$
$a = \dfrac{\Delta F}{m} = \dfrac{80\,lb}{20\,slug} = 4.0\,ft/s^2$ uphill

4-8

A 58 Newton force is acting on the 18 kg block as shown. The coefficient of kinetic friction between the block and the floor is 0.05.
(a) Calculate the force exerted by the floor on the block?
(b) What is the friction force exerted on the block.
(c) Calculate the resultant horizontal force acting on the block.
(d) What is the resultant vertical force acting on the block?
(e) What is the acceleration of the block?

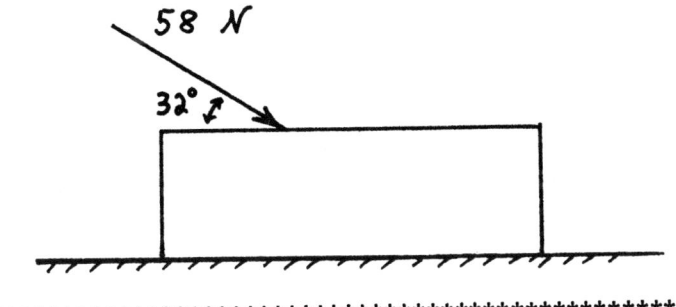

(a) $+\uparrow \Sigma F_y = 0 \qquad F_N - 18(9.8) - 58 \sin 32 = 0 \qquad \underline{\underline{F_N = 206.7 \text{ N} \uparrow}}$

(b) $f_k = \mu_k F_N = 0.05(206.7) = \underline{\underline{10.3 \text{ N} \leftarrow}}$

(c) $\xrightarrow{+} \Sigma F_x = 58 \cos 32 - 10.3 = \underline{\underline{38.9 \text{ N} \rightarrow}}$

(d) Since the block is moving in the x-direction $\underline{\underline{\Sigma F_y = 0}}$

(e) $\Sigma F_x = ma \quad$ and $\quad a = \dfrac{38.9}{18} = \underline{\underline{2.16 \text{ m/s}^2 \rightarrow}}$

4-9

A 20-kg object is acted upon by a force whose X- and Y-components vary with time by the relationship:

$F_x = 20 + 4t$, and $F_y = 40 - 6t^2$, where F is in N and t is in s.

a. Find the resultant force on the object at t = 2 s.
b. Find the expression for the X- and Y-component's of the object's acceleration as a function of time.
c. Find the resultant acceleration of the object at t = 3 s.
d. If the object is initially at rest, find the X- and Y-components of its velocity after 2 s, and find its resultant velocity at t = 2 s.

****

a. $F_x(2) = 20 + 4(2) = 28N$; $F_y(2) = 40 - 6(2)^2 = 16N$
$F = \sqrt{(F_x)^2 + (F_y)^2} = \sqrt{(28N)^2 + (16N)^2} = 32.3 N$
$\theta = \text{Tan}^{-1} \frac{16N}{28N} = 29.7°$
$\therefore F(2s) = 32.3 N$ @ $29.7°$ above +X-axis.

b. $a = \frac{F}{m}$; $\therefore a_x = \frac{F_x}{m} = \frac{20 + 4t}{20} = 1 + 0.2t$, where a is in m/s² and t is in s;

and $a_y = \frac{F_y}{m} = \frac{40 - 6t^2}{20} = 2 - 0.3t^2$, where a is in m/s² and t is in s.

c. $a_x(3) = 1 + 0.2(3) = 1.6$ m/s² ; $a_y(3) = 2 - 0.3(3)^2 = -0.7$ m/s²
$a = \sqrt{(a_x)^2 + (a_y)^2} = \sqrt{(1.6 m/s²)^2 + (-0.7 m/s²)^2} = 1.75$ m/s²
$\theta = \text{Tan}^{-1} \frac{-0.7 m/s²}{1.6 m/s²} = -23.6°$
$\therefore a(3s) = 1.75$ m/s² @ $23.6°$ below +X-axis.

d. $v_x = \int_0^{2s} (1 + 0.2t) dt = t + 0.1t^2 \Big|_0^2 = (2 + 0.4)$ m/s
$= 2.4$ m/s

$v_y = \int_0^{2s} (2 - 0.3t^2) dt = 2t - 0.1t^3 \Big|_0^2 = (4 - 0.8)$ m/s
$= 3.2$ m/s

$v = \sqrt{v_x^2 + v_y^2} = \sqrt{(2.4 m/s)^2 + (3.2 m/s)^2} = 4.0$ m/s
$\theta = \text{Tan}^{-1} \frac{3.2 m/s}{2.4 m/s} = 53.1°$
$\therefore v(2s) = 4.0$ m/s @ $53.1°$ above +X-axis.

4-10

Three blocks are connected by cords and are pulled to the right by a force T_3 as shown below. There is no friction and the entire system is accelerating at .10 m/sec². The blocks have weights of 9.8, 19.8 and 29.4 newtons as marked. Take g as 9.8 m/sec². Calculate the tension in each cord.

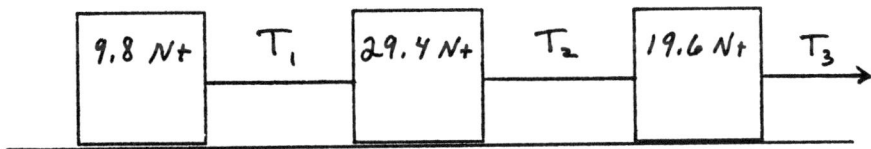

****

Force T_3 is the net force on the system along the plane.

$$T_3 = M_{TOT}\, a = \left(\frac{9.8 + 29.4 + 19.6}{9.8 \text{ m/sec}^2}\right)(.10) \text{ m/sec}^2 = \underline{.60 \text{ Nt}}$$

Apply Newton's 2nd law to the right block.

$$T_3 - T_2 = \left(\frac{19.6 \text{ Nt}}{9.8 \text{ m/s}^2}\right) a \quad \text{so} \quad T_2 = T_3 - 2.0\, a$$

$$T_2 = .60 \text{ Nt} - 2.0(.10) \text{ Nt} = \underline{.40 \text{ Nt}}$$

Finally, apply F=ma to the left block.

$$T_1 = \left(\frac{9.8}{9.8}\right) a = \underline{.10 \text{ Nt}}$$

As a cross-check, examine the center block:

$$T_2 - T_1 = \left(\frac{29.4}{9.8}\right) a$$

$$.40 - .10 = (3.0)(.10) \quad OK$$

4-11

An elevator cab, including its passenger, weighs 2,000 pounds. The cable supporting the elevator cab has a tension of 2,400 pounds during a portion of a trip. At this time, (a) what are the magnitude and direction of the elevator's acceleration, and (b) what is the apparent weight of the passenger if he really weighs 180 pounds?

****

Free-body diagram of elevator cab:

↑ 2400 lbs
↓ 2000 lbs

$F = ma \Rightarrow a = \dfrac{F}{m} = \dfrac{F(g)}{(W)}$

$m = \dfrac{W}{g}$

$a = \dfrac{(2400\,lbs - 2000\,lbs)(32\,\frac{ft}{sec^2})}{(2000\,lbs)}$

$a = 6.4\,\dfrac{ft}{sec^2}$, upward

Free-body diagram of passenger:

↑ A = ? (apparent wt.)
↓ W = 180 lbs

$F = ma \Rightarrow A - W = ma = \dfrac{W}{g}a$

$A = W + W\left(\dfrac{a}{g}\right)$

$A = (180\,lbs)\left[1 + \dfrac{6.4}{32}\right] = 216\,lbs$

4-12

A 10 N block is at rest on a 40 N block which in turn rests on a frictionless table. The coefficient of static friction between the 10 N block and the 40 N block is .4 and the coefficient of kinetic friction is .2. A force of 25 N acts on the 40 N block as shown. (a) What is the resulting acceleration of the 40 N block? (b) What is the resulting acceleration of the 10 N block? (c) What is the resulting acceleration of the center of mass?

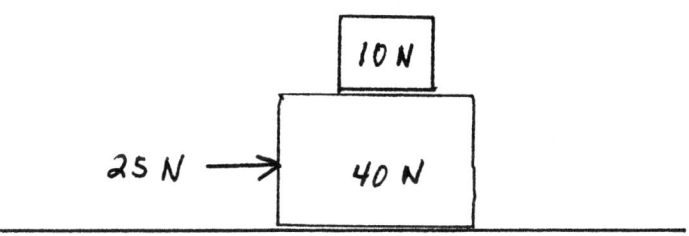

****

a) Maximum force that 40N block can exert on 10N block is $\mu_s N_{10} = .4(10N) = 4N$
IF 10N block does not slip on 40N block, the acceleration of the system is

$$a = \frac{F}{m} = \frac{25N}{\frac{50N}{9.8 m/s^2}} = (.5)(9.8 m/s^2) = 4.9 m/s^2$$

This would require a force on the 10N block of

$$F = ma = \frac{10N}{9.8 m/s^2}(4.9 m/s^2) = 5N$$

This is greater than 4N, so the 10N block slips and we must use coefficient of kinetic friction. Net force on 40N block is

$$F_{net} = 25N - (.2)(10N) = \frac{40N}{9.8 m/s^2}(a_{40})$$

$$a_{40} = \frac{23N}{40N}(9.8 m/s^2) = 5.64 m/s^2$$

b) $a_{10N} = \frac{F}{m_{10}} = \frac{2N}{10N}(9.8 m/s^2) = 1.96 m/s^2$

c) $a_{cm} = \frac{F_{ext}}{m_{TOTAL}} = \frac{25N}{50N}(9.8 m/s^2) = 4.9 m/s^2$

4-13

A small bug is placed between two blocks of masses m_1 and m_2 ($m_1 > m_2$) on a frictionless table. A horizontal force F is applied to one of the blocks as shown. Show that the bug would have a greater chance of survival when the force is applied to the block of larger mass.

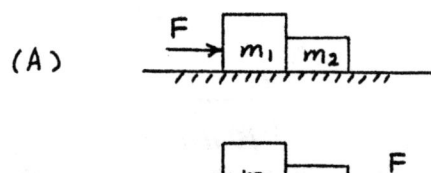

****

Let the force of contact between the blocks be F'. When the horizontal force F is applied to m_1 (case A),
$$F = (m_1 + m_2) a.$$
Also, $F - F' = m_1 a$. Therefore $F' = F - m_1 \dfrac{F}{m_1 + m_2}$
$= F\left(\dfrac{m_2}{m_1 + m_2}\right)$. On the other hand, when F is applied to m_2 (case B),
$$F = (m_1 + m_2) a,$$
but also $F - F' = m_2 a$. Therefore, $F' = F\left(\dfrac{m_1}{m_1 + m_2}\right)$.
Since $m_1 > m_2$, $F\left(\dfrac{m_2}{m_1 + m_2}\right) < F\left(\dfrac{m_1}{m_1 + m_2}\right)$, that is, the bug would experience a smaller force between the blocks in the case A and it would have a better chance of survival.

4-14

A 5.0 kg block is pushed along the ceiling by a constant applied force of 100 N which acts at an angle of 36.87° from the vertical as shown in the figure below. If the block accelerates to the right at 8.0 m/s², find the coefficient of kinetic friction between the block and the ceiling. (Use 9.8 m/s² ≅ 10 m/s².)

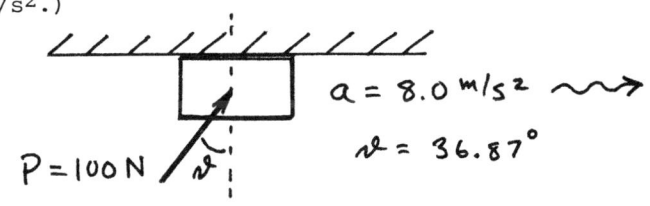

****

First identify the forces and resolve them into x & y components using x-axis parallel to the ceiling.

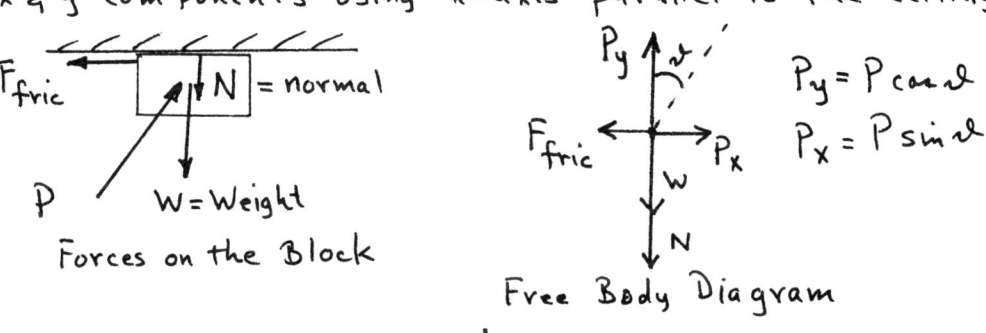

x Forces	y Forces
Newton II: $\sum F_x = ma_x$	$\sum F_y = ma_y$
$P\sin\alpha - F_{fric} = ma_x$	$P\cos\alpha - W - N = 0$
$P\sin\alpha - \mu N = ma_x$	$N = \boxed{P\cos\alpha - W}$

substitute

$$P\sin\alpha - \mu(P\cos\alpha - W) = ma_x$$

$$\mu = \frac{P\sin\alpha - ma_x}{P\cos\alpha - W} = \frac{60 - 40}{80 - 50} = \underline{0.67}$$

4-15

Three masses are connected by a light string as shown below. The string connecting masses two and three passes over a frictionless peg. Using the mass values as shown in the figure, find:

(a) the acceleration of the system (including direction, and

(b) the tension in each of the two strings.

Use $9.8 \text{ m/s}^2 \doteq 10 \text{ m/s}^2$.

$m_1 = 3.0 \text{ kg}$
$m_2 = 3.0 \text{ kg}$
$m_3 = 4.0 \text{ kg}$

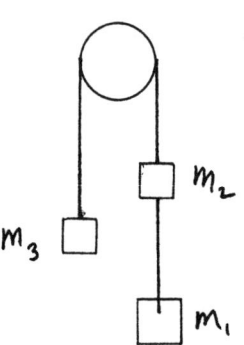

****

Take forces which move the system clockwise as positive. (i.e. force which tend to move m_1 and m_2 down or m_3 up are positive.)

(a) The total force tending to move the system is $+W_1 + W_2 - W_3$; the total mass being accelerated is $m_1 + m_2 + m_3$. Thus Newton's 2nd law gives

$$\sum_{system} F = (m_{system}) a$$

$$W_1 + W_2 - W_3 = (m_1 + m_2 + m_3) a$$

$$a = \left(\frac{m_1 + m_2 - m_3}{m_1 + m_2 + m_3}\right) g = \frac{2}{10} g$$

$$\underline{\underline{a = 2 \text{ m/s}^2}}$$

(b) Analysing m_1 only:

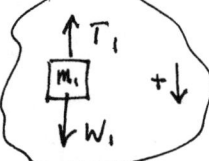

$W_1 - T_1 = m_1 a$

$T_1 = W_1 - m_1 a$
  $= 30 \text{ N} - (3 \text{ kg})(2 \text{ m/s}) = \underline{\underline{24 \text{ N}}} = T_1$

for m_3: $T_2 - W_3 = m_3 a$

$T_2 = W_3 + m_3 a$
  $= 40 \text{ N} + (4 \text{ kg})(2 \text{ m/s}) = \underline{\underline{48 \text{ N}}}$

4-16

A block weighing 128 N is to be dragged down a 36.87° incline at a constant velocity of 0.02 m/s. What horizontal force, P, is necessary to accomplish this? Take the coefficient of kinetic friction between the block and the plane to be 0.8.

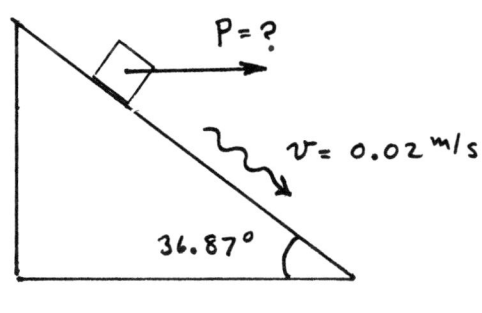

***

Forces on the block are: Weight, Normal, Friction, Applied

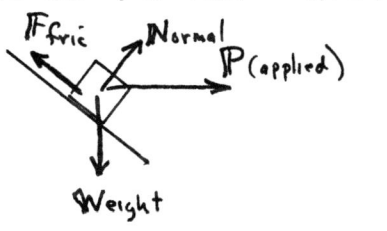 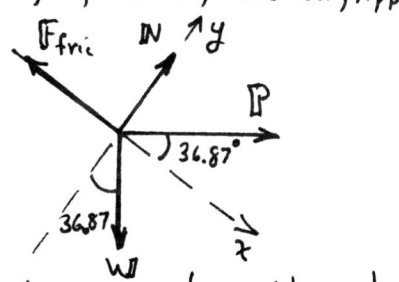

Resolving P and W into components along the plane (x) and perpendicular to the plane (y) we get

$$P_x = P\cos 36.87° = 0.8P \; ; \; P_y = P\sin 36.87° = 0.6P$$
$$W_x = W\sin 36.87° = 0.6W \; ; \; W_y = W\cos 36.87° = 0.8W$$

Using Newton's 2nd law for x & y directions

$\Sigma F_x = ma_x$	$\Sigma F_y = ma_y$
$P_x + W_x - F_{fric} = ma_x$	$N + P_y - W_y = ma_y$
$0.8P + 0.6W - \mu N = 0$	$N + 0.6P - 0.8W = 0$
(note v = constant $\Rightarrow a_x = 0$)	$N = 0.8W - 0.6P$

substitute this expression for N

$$0.8P + 0.6W - \mu(0.8W - 0.6P) = 0 \quad \text{giving}$$

$$P = \frac{\mu(0.8)W - 0.6W}{\mu(0.6) + 0.8} = \underline{4.0 \text{ N}}$$

4-17

A 5 kg block is given a continuous horizontal push so that it slides on a horizontal frictionless surface in a straight line with a constant acceleration of 4.2 m/s² in the direction in which it is moving.
a) Draw a free-body diagram showing all forces that act on the block. Include the acceleration vector in this diagram.
b) What is the magnitude of the normal force on the block due to the surface?
c) What is the magnitude of the push that is acting on the block?

****

a) Free body diagram

$a = 4.2$ m/s²

[Diagram: 5 kg block with push P on left, weight W = 5(9.8) = 49 N down, normal N up]

b) $\Sigma F_{vert} = 0$: $-49 + N = 0 \Rightarrow N = 49.0$ N

c) $\Sigma F_{hor} = Ma_{hor}$: $P = 5(4.2) = 21.0$ N

CONSTANT ACCELERATION

4-18

YOU WANT TO DETERMINE THE COEFFICIENTS OF STATIC AND DYNAMIC FRICTION BETWEEN YOUR SNEAKERS AND YOUR MOTHER'S NEW TABLE. YOU STAND ON ONE END OF THE TABLE AND YOUR FRIEND RAISES THAT END SLOWLY. WHEN THE TABLE IS AT AN ANGLE OF 30 DEGREES, YOU START TO SLIP, MOVING 4.0 METERS IN 4.0 SECONDS. WHAT ARE THE COEFFICIENTS OF FRICTION?

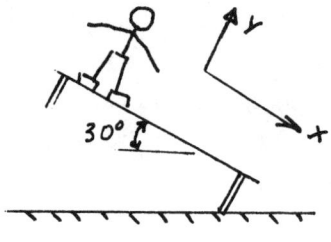

****

When the table is at 30 degrees, you are on the "verge of slipping". Therefore, the friction force is equal to the normal force times the coefficient of static friction. Acceleration is zero.

1) $\Sigma F_x = 0 \quad \therefore W \sin 30° = F = \mu_s N$

2) $\therefore \mu_s = W\sin 30°/N$

3) $\Sigma F_y = 0 \quad \therefore N = W\cos 30°$

4) From 2 & 3, $\mu_s = \dfrac{W\sin 30°}{W\cos 30°} = \tan 30° = .58$

Once you begin to slip, the coefficient of friction decreases from the static value to the kinetic value. The weight force component in the direction of motion is larger than the opposing friction force and you accelerate.

Using Newton's second law

5) $\Sigma F_x = Ma_x \quad \therefore W\sin 30° - \mu_k N = Ma_x$

6) $\Sigma F_y = 0 \quad \therefore N = W\cos 30°$

7) from 5 & 6 $\mu_k = \dfrac{W\sin 30° - Ma_x}{W\cos 30°}$

$= \dfrac{.5 Mg - Ma_x}{.867\, Mg}$

Need to find the acceleration; therefore using given information

8) $x = \frac{1}{2} a_x t^2$

9) $\therefore a = \dfrac{2x}{t^2} = \dfrac{2 \times 4m}{(4s)^2} = .5\, m/s^2$

Plugging value for acceleration into equation 3, and solving

10) $\mu_k = \dfrac{.5 \times 9.8\, m/s^2 - .5\, m/s^2}{.867 \times 9.8\, m/s^2} = .52$

4-19

At the instant a traffic light turns green, a waiting automobile starts forward with a constant acceleration of 6.0 ft/s^2. At the same instant a truck, traveling at a constant speed of 30 ft/s passes the automobile.
a. At what point does the auto overtake the truck?
b. What is the speed of the auto then?

***

a) For the auto, we write
$$x_A = v_0 t + \tfrac{1}{2} a t^2$$
$$= \tfrac{1}{2} \times 6 \times t^2 = 3t^2$$

For the truck
$$x_T = vt = 30t$$

Setting $x_A = x_T$
$$3t^2 = 30t$$
$$t = 10 s$$

Substituting into either x_T or x_A
$$x_A = x_T = 300 \text{ f}$$

b)
$$v_A = v_0 + at$$
$$v_A = 0 + 6 \times 10$$
$$v_A = 60 \text{ ft/s}$$

4-20

A railroad flatcar is loaded with crates having a coefficient of static friction of 0.50 with the floor. If the train is moving at 88 feet/second before it uniformly decelerates to a stop without causing the crates to slide, the minimum stopping distance for the train's flatcar is what?

****

$v_0 = 88$ feet/second
$v = 0$
$\mu = 0.50$
$a = $ constant < 0

$$v^2 = v_0^2 + 2as$$

$$2as = v^2 - v_0^2$$

$$s = \frac{v^2 - v_0^2}{2a}$$

net $F = ma$
$F_f = -\mu N = -\mu mg$
$ma = -\mu mg$
$a = -\mu g$

$$s = \frac{v^2 - v_0^2}{2(-\mu g)}$$

$$s = \frac{0 - v_0^2}{-2\mu g}$$

$$s = \frac{v_0^2}{2\mu g}$$

$$s = \frac{(88 \text{ feet/second})^2}{2(0.50)(32 \text{ feet/sec}^2)}$$

$$\boxed{s = 242 \text{ feet}}$$

4-21

Consider the following system of particles when t = 0 seconds:
m_1 = 2 kg, x_1 = 0, y_1 = 3m; m_2 = 4 kg, x_2 = -2, y_2 = 0;
m_3 = 6 kg, x_3 = -3, y_3 = -3. Suppose that the velocities of
these particles are given by:

$\vec{v}_1 = (4 + 2t)\ \vec{i}$;
$\vec{v}_2 = 5\vec{j}$; and
$\vec{v}_3 = (6 + 3t)\ \vec{j}$.

Find the location of the center of mass of the system when
t = 4 seconds.

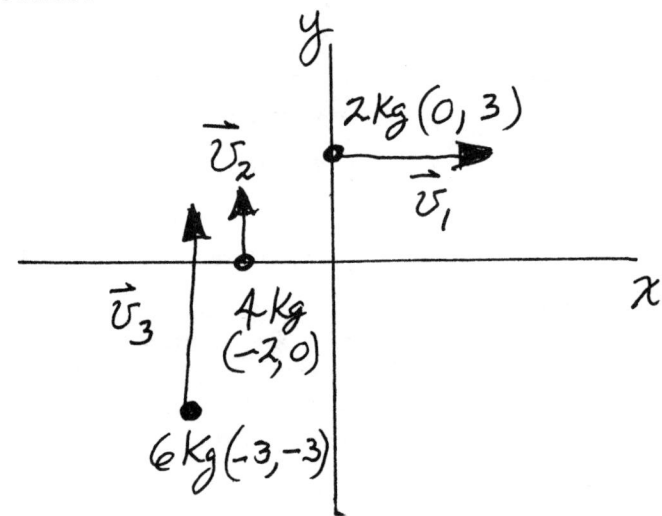

****

Method I: Note first that $\vec{a}_1 = \dfrac{d\vec{v}_1}{dt} = 2\vec{i}$

then
$$X_1[t=4] = X_1[t=0] + v_{x_1}[t=0]\Delta t + \tfrac{1}{2}a_{1x}(\Delta t)^2$$
$$= 0 + (4)(4) + \tfrac{1}{2}(2)(4)^2 = 32$$

Similarly $X_2[t=4] = -2$, $X_3[t=4] = -3$

and $X_{cm}[t=4] = \dfrac{\sum_{i=1}^{3} m_i X_i}{\sum_{i=1}^{3} m_i} = \dfrac{(2)(32)+4(-2)+(6)(-3)}{12}$

$X_{cm}[t=4] = 3\tfrac{1}{6}$, likewise $y_{cm}[t=4] = 29\tfrac{2}{3}$

because $y_{cm}[t=4] = \dfrac{(2)(3)+(4)(20)+(6)(45)}{12}$

Method 2:

First find $\vec{R}_{cm}[t=0] = x_{cm}[t=0]\vec{i} + y_{cm}[t=0]\vec{j}$

which gives $\vec{R}_{cm}[t=0] = -2\tfrac{1}{6}\vec{i} - 1\vec{j}$

Next note that $\vec{v}_{cm} = \dfrac{\sum m_i \vec{v}_i}{\sum m_i}$

So that $\vec{v}_{cm} = \dfrac{(2)(4+2t)\vec{i} + (4)(5\vec{j}) + (6)(6+3t)\vec{j}}{12}$

$= \dfrac{(8+4t)\vec{i} + (56+18t)\vec{j}}{12}$

and $\vec{v}_{cm}[t=0] = \dfrac{8\vec{i} + 56\vec{j}}{12}$

Now $\vec{a}_{cm} = \dfrac{d\vec{v}_{cm}}{dt} = \dfrac{\sum \vec{F}_i}{\sum m_i} = \dfrac{4\vec{i}+18\vec{j}}{12}$

and $\vec{R}_{cm}[t=4] = \vec{R}_{cm}[t=0] + \vec{v}_{cm}[t=0]\Delta t + \tfrac{1}{2}\vec{a}_{cm}(\Delta t)^2$

$= (-2\tfrac{1}{6}\vec{i} - \vec{j}) + (2\tfrac{4}{6}\vec{i} + 18\tfrac{4}{6}\vec{j}) + (2\tfrac{4}{6}\vec{i} + 12\vec{j})$

$= 3\tfrac{1}{6}\vec{i} + 29\tfrac{2}{3}\vec{j}$

4-22

A 6 Kg. block slides down an incline plane pulled by a weightless chord attached to a 4 Kg hanging mass by means of a pulley. There is no friction on the incline or in the pulley. The incline plane makes an angle of 20° to the horizontal. What is the acceleration of the two blocks and the tension in the chord?

****

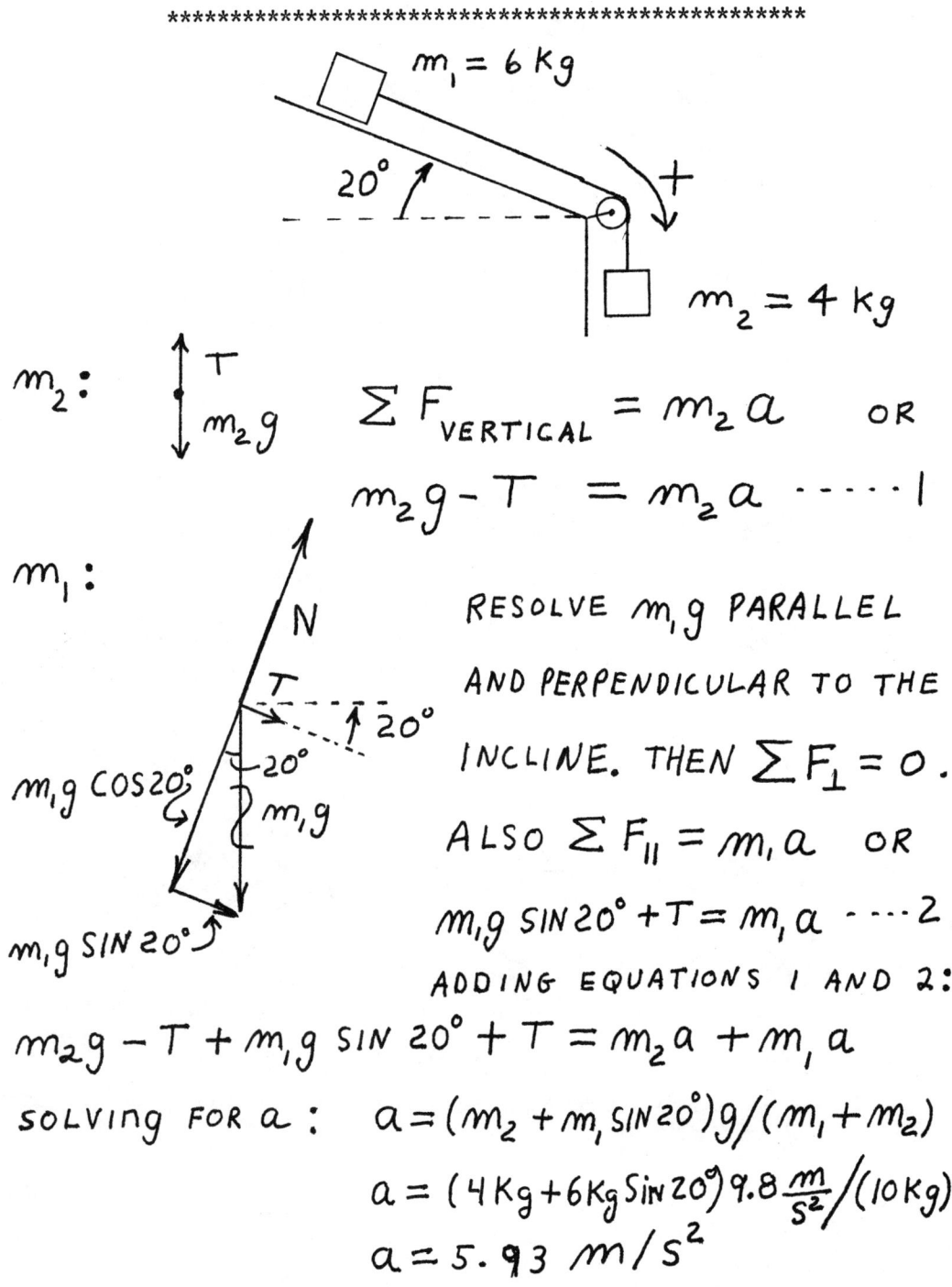

m_2:

$\sum F_{VERTICAL} = m_2 a$ OR

$m_2 g - T = m_2 a$ ·····1

m_1:

RESOLVE $m_1 g$ PARALLEL AND PERPENDICULAR TO THE INCLINE. THEN $\sum F_\perp = 0$.

ALSO $\sum F_\parallel = m_1 a$ OR

$m_1 g \sin 20° + T = m_1 a$ ····2.

ADDING EQUATIONS 1 AND 2:

$m_2 g - T + m_1 g \sin 20° + T = m_2 a + m_1 a$

SOLVING FOR a: $a = (m_2 + m_1 \sin 20°)g/(m_1 + m_2)$

$a = (4 Kg + 6 Kg \sin 20°) 9.8 \frac{m}{s^2} / (10 Kg)$

$a = 5.93 \, m/s^2$

SOLVING EQUATION 1 FOR T GIVES:

$$T = m_2(g-a) = 4 \text{ Kg}(3.87 \text{ m/s}^2) = 15.5 \text{ N}$$

SUBSTITUTE IN EQUATION 2 TO CHECK:

$$6 \text{ Kg } 9.8 \text{ m/s}^2 \sin 20° + 15. \text{ N} \stackrel{?}{=} 6 \text{ Kg } 5.93 \text{ m/s}^2$$

OR $35.6 \cong 35.6$ N WHICH IS EQUAL TO THE ACCURACY OF THE CALCULATION, ABOUT 1%.

4-23

A body moves on a straight path with constant acceleration $a = -8$ m/sec^2. At $t = 0$, its position is $x = 0$ and its velocity is $v = 10$ m/sec.
(a) Compute the time and location at which $v = 0$.
(b) Compute the velocity when $x = -100$ m.

****

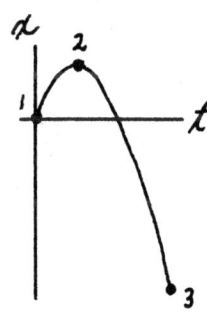

Position vs. time $(x_1 = t_1 = 0)$

(a) $v_2 = v_1 + a t_2$

Set $v_2 = 0$, $v_1 = 10$, $a = -8$

Then $t_2 = \dfrac{v_1}{-a} = \underline{1.25 \text{ sec}}$

$x_2 = v_1 t_2 + \tfrac{1}{2} a t_2^2 = \underline{6.25 \text{ m}}$

(b) $v_3^2 = v_1^2 + 2 a x_3$ Set $v_1 = 10$, $a = -8$, $x_3 = -100$

Then $v_3^2 = 1700$, $v_3 = \pm 41.2$ m/sec.

From the slope of the above curve at point 3, clearly $v_3 = \underline{-41.2 \text{ m/sec}}$

4-24

A parachutist, during an exhibition, steps off a tall building and free-falls vertically from rest for a distance of 44.1 meters (no air resistance). He then opens his parachute and decelerates at 2 meters/sec^2. He reaches the ground with a speed of 3.4 meters/sec. (a) How long is the parachutist in the air? (b) How tall is the building?

****

(a) Choose the downwards direction as positive.

$$y = V_0 t + \tfrac{1}{2} a t^2 \qquad 44.1 = (0)(t) + \tfrac{1}{2}(9.8) t^2$$

$t = 3$ sec before the parachute opens.

$$V = V_0 + a t \qquad V = (0) + (9.8)(3) = 29.4 \text{ m/sec}$$

Now use this as initial velocity for second part of fall.

$$V = V_0 + a t \qquad 3.4 = 29.4 + (-2) t \qquad t = 13 \text{ sec}$$

Total time in air = $3 + 13 = 16$ sec.

(b) $y = V_0 t + \tfrac{1}{2} a t^2 = (29.4)(13) + \tfrac{1}{2}(-2)(13)^2$

$= 213.2$ m = distance he fell with parachute open.

$44.1 + 213.2 = 257.3$ m = height of building.

4-25

A mass (m_1 = 4 kg) slides on a rough horizontal surface with a kinetic coefficient of friction μ_k = 0.7. A second mass (m_2 = 3 kg) is attached to m_1 by a light string draped over a light, frictionless pulley. A force (F) is applied to m_1 causing the system to move with a constant speed to the left. (a) What is the tension in the string? (b) What is the magnitude of F? (c) If F is removed, what coefficient of static friction would be required between the horizontal surface and m_1 to keep the system from moving?

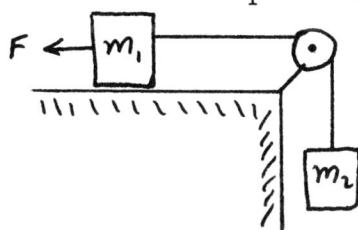

****

(a) Draw a free-body diagram for M_2

Since there is no acceleration
$$\Sigma F_2 = 0$$
$$T - m_2 g = 0$$
$$T = m_2 g = (3)(9.8) = \underline{29.4 \text{ N}}$$

(b) Draw a free-body diagram for M_1

T = tension in string
f = frictional force = $\mu_k N$
N = Normal force

vertically: $\Sigma F_1 = 0$
$$N - m_1 g = 0$$
$$N = M_1 g = 39.2 \text{ N}$$

horizontally: $\Sigma F_1 = 0$
$$F - T - f = 0$$
$$F = T + f$$
$$= 29.4 + (0.7)(39.2)$$
$$F = \underline{56.84 \text{ N}}$$

(c) Draw new diagram for M_1

f_s = static frictional force
$$f_s \leq \mu_s N \leq \mu_s M_1 g$$

horizontally: $\Sigma F_1 = 0$
$$f_s - T = 0$$
$$f_s = T$$

$$\mu_s \geq \frac{T}{Mg}$$
$$\underline{\mu_s \geq 0.75}$$

4-26

A 1-kg mass is <u>dropped</u> from a height of 420 meters. A second 1-kg mass is <u>thrown</u> vertically downward 5 seconds after the first mass is released. Determine the initial velocity of the second mass if the two masses strike the ground at the same time. Neglect air resistance.

****

<u>FIRST</u>: figure out how long it takes for the dropped mass to hit the ground:

constant acceleration $= g$; $v_0 = 0$ (dropped from rest)
$y_0 = 0$; $y = 420$ m   (taking $\oplus$ down)

use $y = y_0 + v_{0y} t + \frac{1}{2} a_y t^2$
   $\underbrace{420m}\ \underbrace{0}\ \underbrace{0}\ \ \ \ \underbrace{g}$

$$420 \text{ m} = \frac{1}{2} \cdot (9.8 \text{ m/s}^2) t^2 \Rightarrow \underline{t = 9.26 \text{ sec}}$$

The second mass was thrown down 5 seconds later, and hits the ground at the same time $\Rightarrow$ it took $(9.26 - 5) = 4.26$ seconds to fall. Using the same formula as before (be sure to use the same acceleration! just because the second mass had a non-zero initial <u>speed</u> doesn't mean its <u>acceleration</u> is different!)

$$y' = y_0' + v_{0y}' t' + \frac{1}{2} a_y t'^2$$

$y' = 420$ m, $y_0' = 0$, $t' = 4.26$ sec, $a_y = g = 9.8$ m/s^2

Solve for $\boxed{v_{0y}' = 77.8 \text{ m/s}.}$

4-27

A man stands on a bridge crossing a river as a long barge passes underneath. He drops a stone from the bridge at a height of 144 feet from the barge deck. He drops a second stone 10 seconds after he released the first and notes that it lands 35 feet from the spot where the first stone landed on the barge. Assume that the barge is accelerating at a constant rate of 0.10 feet/second² and answer the following questions:

a) How long did it take for the first stone to hit the barge?
b) What was the instantaneous speed of the barge when the first stone landed on the barge?
c) What was the instantaneous speed of the barge when the second stone landed on the barge?

****

a) $y = \frac{1}{2}gt^2$ so $t = \sqrt{\frac{2y}{g}} = \sqrt{\frac{2(144)}{32}} = \sqrt{9} = \underline{3.0 \text{ sec}}$

b) Let x = distance barge moves in 10 seconds.

$$x = x_0 + V_{ox}t + \frac{1}{2}a_x t^2$$
$$35 \text{ ft} = 0 + V_{ox}(10 \text{ sec}) + \frac{1}{2}(.10 \text{ ft/sec}^2)(10 \text{ sec})^2$$
$$35 \text{ ft} = 10 V_{ox} + 5 \text{ ft}$$
$$V_{ox} = \frac{30 \text{ ft}}{10 \text{ sec}} = \underline{3.0 \text{ ft/sec}}$$

c) Let v = speed of barge when second stone hits it.

$$V = V_{ox} + at = 3.0 \text{ ft/sec} + (.10)(10) \text{ ft/sec.}$$
$$= \underline{4.0 \text{ ft/sec}}$$

Or, using the horizontal distance instead of the time,

$$v^2 = V_{ox}^2 + 2a(x - x_0) = (3.0)^2 + 2(.10)(35)$$
$$= 9 + 7 = 16 \text{ ft}^2/\text{sec}^2$$

$v = \underline{4.0 \text{ ft/sec}}$. Note that the time of fall for the stones did not enter into the solution of b or c.

4-28

A 4 kg block is connected to a 1 kg block by a massless string over a frictionless pulley. The 4 kg block slides on a frictionless horizontal surface as shown. (a) What is the acceleration of the system and the tension in the string when the 1 kg block is released? (b) If the 1 kg block is .5 m above the floor, how long does it take to hit the floor after it is released?

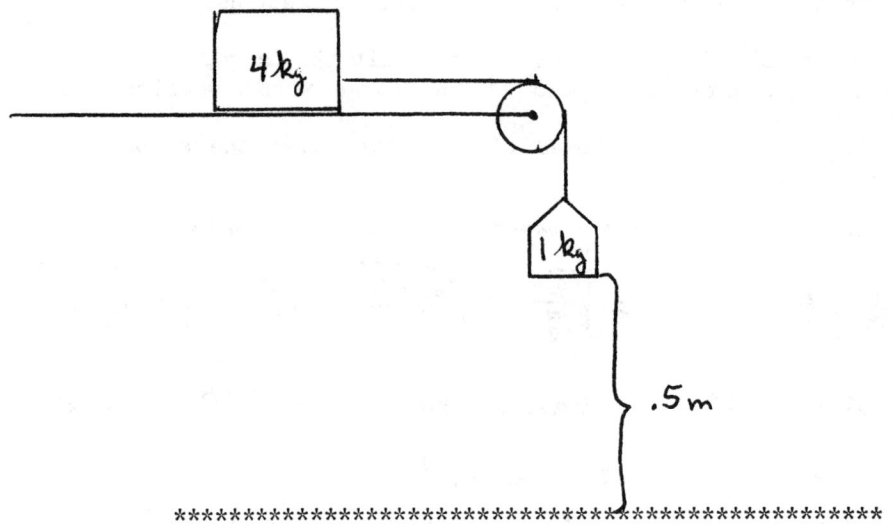

***

a) For 4 kg
$\Sigma F_y = 0$
$\Sigma F_x = T = 4\text{ kg } a$

For 1 kg
$\Sigma F_x = 0$
$\Sigma F_y = T - (1\text{ kg})(9.8\text{ m/s}^2) = (1\text{ kg})(-a)$

$T = 9.8\text{ N} - (1\text{ kg}) a$

$(4\text{ kg}) a = 9.8\text{ N} - (1\text{ kg}) a$
$a = 1.96 \text{ m/s}^2$

$T = 9.8\text{ N} - (1\text{ kg})(1.96\text{ m/s}^2) = 7.84\text{ N}$

b) $y - y_0 = v_{y0} t + \frac{1}{2} a_y t^2$

$0 - .5\text{ m} = 0 + \frac{1}{2}(-1.96\text{ m/s}^2) t^2$

$t = \sqrt{\dfrac{2(-.5\text{ m})}{-1.96\text{ m/s}^2}} = .71\text{ s}$

4-29

Starting at t=0 and for the next four seconds the time and position coordinates of an air track glider, moving in a straight line, were measured to be (t sec, x cm) = (0, 5), (0.1, 10), (0.2, 19), (0.3, 32), (0.4, 49). Find the constant acceleration and the initial velocity at t=0.

****

USE $\Delta x = x_f - x_0$, $v = \Delta x / \Delta t$, $\Delta v = v_f - v_i$, and $a = \Delta v / \Delta t$

t (SEC)	x (cm)	Δx (CM)	v (CM/SEC)	Δv (CM/SEC)	a (CM/S²)
0	5				
		5	50		
0.1	10			40	400
		9	90		
0.2	19			40	400
		13	130		
0.3	32			40	400
		17	170		
0.4	49				

USE $t = 0.1$ SEC AND $x = x_0 + v_0 t + \frac{1}{2} a t^2$ to FIND v_0:

$10 \text{ CM} = 5 \text{ CM} + (0.1 \text{ SEC}) v_0 + \frac{1}{2} 400 \text{ CM/SEC}^2 (0.1 \text{ SEC})^2$

WHICH YIELDS $v_0 = 30$ CM/SEC.

CHECK AT $t = 0.2$ SEC:

$x = 5 \text{ CM} + 30 \frac{\text{CM}}{\text{SEC}} (0.2 \text{ SEC}) + \frac{1}{2} 400 \frac{\text{CM}}{\text{SEC}} (0.2 \text{ SEC})^2$

OR $x = 19$ CM AS ABOVE.

4-30

A 20-kg object accelerates uniformly across a horizontal surface at 4 m/s^2 when a 160-N horizontal force is applied to it.
a. What is the coefficient of kinetic friction between the object and the surface?
b. What force applied 30° above horizontal will cause the object to accelerate at the same rate?

****

a.

$\Delta F = F - f$
$ma = F - \mu N$; $(20 kg)(4 m/s^2) = 160 N - \mu (20 kg)(9.8 m/s^2)$
$\mu = \dfrac{160N - 80N}{196N} = \dfrac{80N}{196N} = 0.41$

b.

$\Delta F = F_h - f$
$ma = F_h - \mu(Wt - F_v)$
$= F\cos 30° - \mu(mg - F\sin 30°)$
$(20 kg)(4 m/s^2) = F(.866) - 0.41[(20 kg)(9.8 m/s^2) - F(0.50)]$
$80N = F(0.866) - 80N + F(0.20)$
$F = \dfrac{80N + 80N}{1.066} = 150 N$

PROJECTILE MOTION

4-31

Wile Y. Coyote pursues the Roadrunner at a constant speed on a level road. The road makes an abrupt turn and Wile Y. goes straight over a vertical cliff 44m above the surface of a river. He lands in the middle of the river 39m from the face of the cliff, which is directly below the point where he left the road. How fast was the coyote running?

****

$x = vt$
$y = \frac{1}{2}gt^2$
$v = \dfrac{x}{t} = \sqrt{\dfrac{g}{2y}} \, x$
$v = \sqrt{\dfrac{9.8}{2 \times 44}} \times 39 = 13 \text{ m/s}$

4-32

An engine breaks away from an airplane travelling in a horizontal direction at 300 (m/s) and at an altitude of 900 (m). The engine has no vertical component of motion at the instant of breakaway, i.e. $V_{oy} = 0$
(a) Neglecting air resistance, find the time required for the engine to strike the ground.
(b) Find the displacement R of the engine along the x-direction (i. e. the range) when it strikes the ground.

****

Remember that the y- and x-component motions are independent of each other.

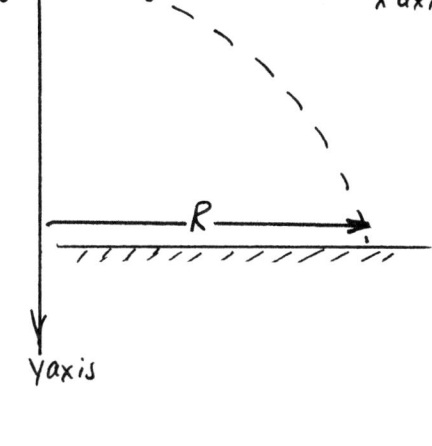

(a) $y = y_o + V_{oy} t + \frac{1}{2} a_y t^2$

$y_o = 0$, $y = 900$ m,

$a_y = +g$ (since we've chosen y-axis downward),

$V_{oy} = 0$.

$y = \frac{1}{2} g t^2$

$t = \sqrt{\frac{2y}{g}} = 14$ s

(b) In this time the engine has moved with constant speed along the x-direction: $V_x = V_{ox} = V_o$

(The initial velocity $\vec{V_o}$ is along x-direction, and the x-component of velocity does not change since $a_x = 0$)

$X - X_o = V_{ox} t$

$V_{ox} = V_o = 300$ m/s

$\therefore R = 300 \text{ (m/s)} \times 14 \text{ (s)} = 4200$ m

4-33

A particle is projected from the ground with a velocity of 100 ft/sec at an angle of 37° above the horizontal. Find the magnitude of its velocity and the angle that it makes to the horizontal after 1.0 seconds of flight time.

****

AT $t=0$ AND $t=t$ ONE HAS:

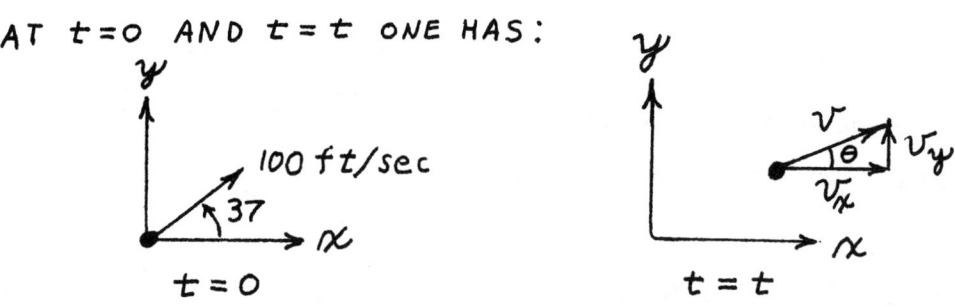

$V_x = V_{xo} = 100 \cos 37 = 80$ ft/sec for all time.

Initially $V_{yo} = 100$ ft/sec $\sin 37 = 60$ ft/sec.

At $t=1$ sec $V_y = V_{yo} - gt = 60$ ft/sec $- 32$ ft/sec^2 1 sec

OR $V_y = 28$ ft/sec. Thus $V = \sqrt{V_x^2 + V_y^2} = \sqrt{(80)^2 + (28)^2}$

OR $V = 85$ ft/sec. $\theta = \tan^{-1} \frac{28}{80} = 19°$

4-34

A tennis ball is shot from a ball-throwing machine an an angle $\theta_o = 30°$ above the horizontal with an initial speed $V_o = 20$ (m/s).
(a) Find the maximum height attained by the ball.
(b) Find the time needed for the ball to reach that height.
(c) Find the range R of the ball.

****

(a) $V_y^2 = V_{oy}^2 + 2a(y-y_o)$

At top of path $V_y = 0$.

So inserting $V_y = 0$ in the equation makes $y - y_o$ correspond to the maximum height.

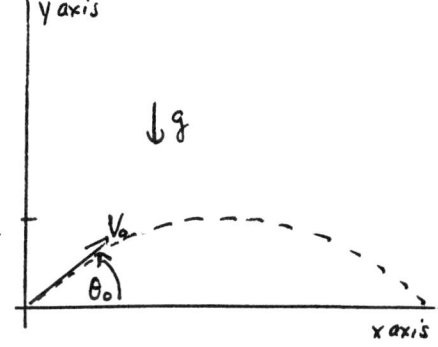

$V_y = 0$
$V_{oy} = V_o \sin\theta_o = 20 \sin 30° = 10$ m/s
$a_y = -g = -9.8$ m/s² (y axis was chosen in the upward direction)

$\therefore 0 = 100 + 2(-9.8)(y-y_o)$

$R = 5.1$ m

(b) $V_y = V_{oy} + a_y t$

When $V_y = 0$, t is time required to reach maximum height.

$\therefore V_y = 0; V_{oy} = V_o \sin\theta_o = 20 \sin 30° = 10$ m/s
$a_y = -g = -9.8$ m/s²

$t = \frac{V_y - V_{oy}}{a_y} = \frac{0-10}{-9.8} = 1.0$ s

(c) Constant speed along x-direction, since $a_x = 0$
$R = x - x_o = V_{ox} T$, where $T = 2t$ (t of part (b)); T is total time ball is in air
$R = V_{ox} T = (V_o \cos\theta_o)(2t) = 20 \times 0.866 \times 2.0 = 35$ m

4-35

A basketball player shoots a ball at a basket 10 ft high from a distance of 24 ft away. The ball leaves the player's hand 8 ft above the floor at an angle of 53° above the horizontal. (a) With what velocity should the basketball player shoot the ball? (b) How long does the ball take to reach the basket? (c) What is the maximum height reached by the ball? (d) At what angle below the horizontal does the ball enter the basket?

****

a) $x - x_0 = 24 \text{ ft} = v_0 \cos 53° \cdot t$

$y - y_0 = 2 \text{ ft} = v_0 \sin 53° \cdot t + \frac{1}{2} a_y t^2$

$2 \text{ ft} = \dfrac{(v_0 \sin 53°)(24 \text{ ft})}{v_0 \cos 53°} - (16 \text{ ft/s}^2)\left(\dfrac{24 \text{ ft}}{v_0 \cos 53°}\right)^2$

$v_0^2 \cos^2 53° (24 \text{ ft} \tan 53 - 2 \text{ ft}) = (16 \text{ ft/s}^2)(24 \text{ ft})^2$

$v_0 = 29.3 \text{ ft/s}$

c) $v_y^2 - v_{y_0}^2 = 2 a_y (y - y_0)$

$y = y_0 + \dfrac{0 - (29.3 \text{ ft} \sin 53°)^2}{2(-32 \text{ ft/s}^2)}$

$= 8 \text{ ft} + 8.55 \text{ ft} = 16.55 \text{ ft}$

b) $t = \dfrac{x - x_0}{v_0 \cos 53°} = \dfrac{24 \text{ ft}}{(29.3 \text{ ft/s})(.6)} = 1.37 \text{ s}$

d) $v_y = v_{y_0} + a_y t = (29.3 \text{ ft/s}) \sin 53° - (32 \text{ ft/s}^2)(1.37 \text{ s})$
$= -20.4 \text{ ft/s}$

$v_x = 29.3 \text{ ft/s} \cos 53° = 17.6 \text{ ft/s} \qquad \tan \theta = \dfrac{-20.4 \text{ ft/s}}{17.6 \text{ ft/s}}$

$\theta = 49.2°$ below horizontal

4-36

You may recall that in the story about Winnie-the-Pooh and the honey tree, Pooh was hanging underneath a balloon and Christopher Robin was trying to get him down by shooting the balloon with a pop gun. Unfortunately, the pop gun had such a low muzzle velocity (100 ft/sec) that Christopher Robin couldn't hit the balloon, even though his aim was perfect. Consider the following sequence of events. Pooh accidently lets go of the balloon string at the instant that Christopher Robin fires his gun. As Pooh falls freely he is hit by the cork from the gun. Using these data and the distances in the sketch, calculate the following:

a) The elapsed time after firing at which Pooh is hit by the cork.
b) The height at which Pooh is hit by the cork.
c) The time between Pooh's encounter with the cork and his landing on the ground.

****

a) Pooh $\quad y = 160' - \overset{0}{v_{oy}} t - \tfrac{1}{2} g t^2$
$\qquad \qquad = 160 - 16 t^2 \text{ feet}$

Cork $\quad y = 0 + V_0 \sin\theta \, t - \tfrac{1}{2} g t^2$
$\qquad \qquad = 100 \text{ ft/s} \left(\tfrac{160}{200}\right) t - 16 t^2$

Equate expressions for y.

$160 - 16 t^2 = 100 (.8) t - 16 t^2$

$t = \dfrac{160}{80} = \underline{2.0 \text{ sec}}$

b) $y = 160 - 16 (2.0)^2 = 160 - 64 = \underline{96 \text{ feet}}$

c) Calculate the time for Pooh to fall 160 feet starting from rest.

$0 = 160' - \tfrac{1}{2} g t_{tot}^2, \quad t_{tot}^2 = \dfrac{320}{32} = 10 \text{ sec}^2$

$t_{tot} = \sqrt{10} \text{ sec} \approx 3.2 \text{ sec}$

$\Delta t = t_{tot} - t = 3.2 \text{ sec} - 2.0 \text{ sec} \approx \underline{1.2 \text{ sec}}$

4-37

Pat Leahy of the New York Jets kicks the football at a 30° angle above the horizontal and just scores a field goal from a point 40 meters in front of the goalposts, whose horizontal bar is 3.5 meters above the ground.

a) How long is the ball in the air before the goal is scored??

b) What was the initial speed of the ball, right after the kick??

****

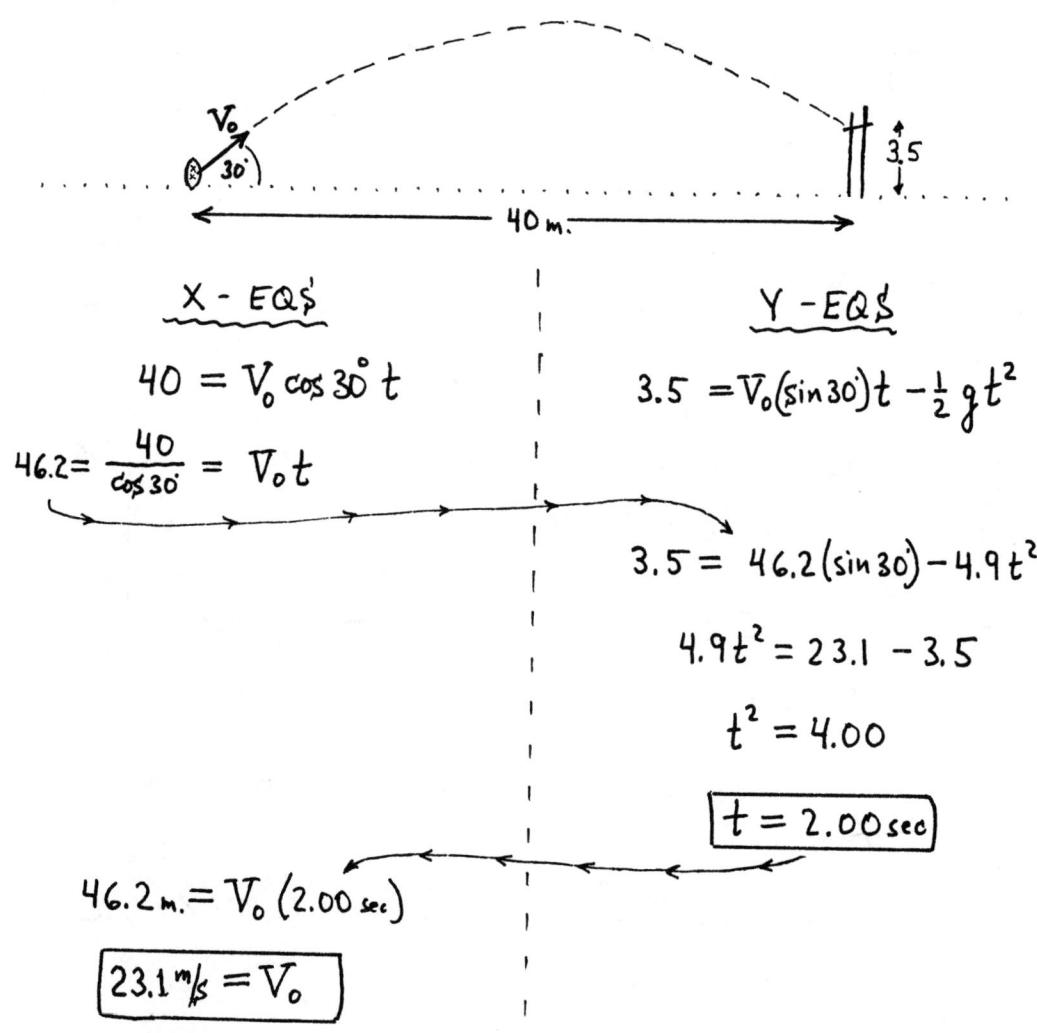

X - EQS

$$40 = V_0 \cos 30° \, t$$

$$46.2 = \frac{40}{\cos 30°} = V_0 t$$

Y - EQS

$$3.5 = V_0 (\sin 30) t - \tfrac{1}{2} g t^2$$

$$3.5 = 46.2 (\sin 30) - 4.9 t^2$$

$$4.9 t^2 = 23.1 - 3.5$$

$$t^2 = 4.00$$

$$\boxed{t = 2.00 \text{ sec}}$$

$$46.2 \text{ m} = V_0 (2.00 \text{ sec})$$

$$\boxed{23.1 \text{ m/s} = V_0}$$

4-38

A student is loading his belongings into a stationwagon in order to depart for college. He temporarily places a box of mass 3.0 kg at the front of the roof of the car, but forgets to remove it before driving away. When the car begins to accelerate, the box slips toward the back of the roof. Let the coefficient of kinetic friction, μ_k, between the box and the roof, be 0.10.

(a) Draw and label a vector diagram showing all forces exerted on the box.
(b) What is the acceleration of the box relative to the ground?
(c) Assuming that the acceleration of the stationwagon is 5.0 m/s^2 and that the length of the roof is 2.5 m, how much time does it take for the box to slide to the back of the roof?
(d) Eventually, the box falls off the back of the roof and hits the ground. Let the height of the roof be 2.0 m. Assuming that the car travels in a straight line along the road (i.e. no turns), what distance away from its initial horizontal position is the box when it hits the ground?

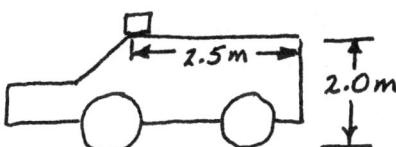

****

(a) [free body diagram: f_k leftward, F_N upward, mg downward]

$F_N = mg$, $f_k = \mu_k F_N = \mu_k mg$

(b) $\vec{F} = \vec{f_k} = m\vec{a}$ So, $ma = \mu_k mg$

$\boxed{a = \mu_k g = 0.98 \text{ m/s}^2}$ DIRECTION IS <u>FORWARD</u>!

(c) $X_{CAR} = \frac{1}{2} a_{CAR} t^2 = 2.5 t^2$ m
$X_{BOX} = \frac{1}{2} a_{BOX} t^2 = 0.49 t^2$ m

BOX IS AT REAR WHEN $X_{CAR} - X_{BOX} = 2.5$ m
$\Rightarrow \boxed{t_1 = 1.1 \text{ s}}$

(d) MOTION OF BOX IS IN 2 PARTS:
PART 1 (ON ROOF): $X_{BOX} = \frac{1}{2} a_{BOX} t_1^2 = 0.61$ m

PART 2 (FALLING AS PROJECTILE): $v_x = a_{BOX} t_1 = 1.1$ m/s
$x = 0.61 + 1.1 t$ (NO X-ACCELERATION)
$y = 2.0 - \frac{1}{2} g t^2$ HITS GROUND WHEN $y = 0$
$\Rightarrow t_2 = 0.64$ s $\Rightarrow \boxed{X = 1.3 \text{ m}}$

4-39

When a balloon whose total height is 5 m is at an altitude of 60 m, it is ascending at a constant rate of 40 m/sec. At this instant a cannon located on the ground below, a distance of 50 m away, fires towards the balloon with a muzzle speed of 200 m/sec and at an azimuth of 55°. Will the ball hit the balloon? If so, what part; if not, by how much will it miss?

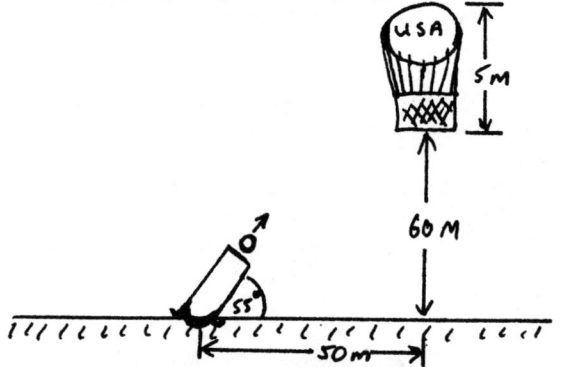

****

What we really need to know is how high both the ball and the balloon will be when the ball has traveled the 50 m horizontal distance:

(i) Find the time it takes the ball to travel the horizontal distance

$$t = \frac{x}{v_x} = \frac{50}{200 \cos 55°} = 0.44 \text{ sec}$$

(ii) Find the height of the ball at this time

$$y = -\tfrac{1}{2} g t^2 + v_{y_0} t$$
$$= -\tfrac{1}{2}(9.8)(0.44)^2 + (200 \sin 55°)(0.44)$$
$$= 71.14 \text{ m}$$

(iii) Find the height of the balloon at this time

$$y = y_0 + v_y t = 60 + (40)(0.44)$$
$$= 77.4 \text{ m}$$

∴ The ball misses the bottom of the balloon by about 6.3 m!

4-40

A ball leaves the bat 1.15 m above the ground at an angle of 35 degrees with the horizontal and traveling at 42 m/s.
(a) When does the ball strike the ground?
(b) What is the maximum height above ground reached by the ball?

****

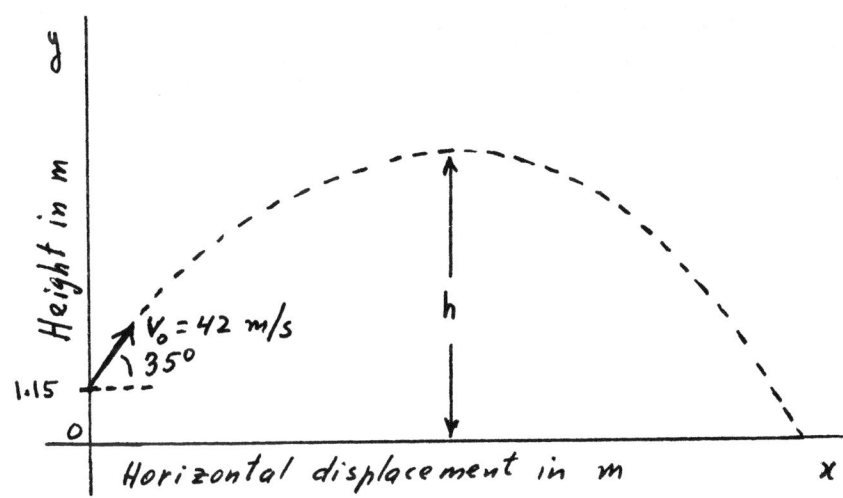

(b) $V_y^2 = V_{y_0}^2 + 2a(y-y_0)$ at $y = h$, $V_y = 0$ and

$+\uparrow$ $0 = V_{y_0}^2 - 2g(h-y_0) = (42 \sin 35)^2 - 2(9.81)(h-1.15)$

and $\underline{h = 30.73 \text{ m}}$

(a) $y = y_0 + V_{0y}t + \frac{1}{2}at^2$ When the ball strikes the ground $y = 0$ and we obtain:

$+\uparrow$ $0 = 1.15 + 42 \sin 35\, t - \frac{1}{2}(9.81)t^2$ or

$-4.905\, t^2 + 24.09\, t + 1.15 = 0$. This is a quadratic equation of the general form $at^2 + bt + c = 0$ and its solution is $t = (-b \pm \sqrt{b^2 - 4ac})/(2a)$ giving $\underline{t = 4.95 \text{ sec.}}$

Note: Part (b) was done first because it was easier.

4-41

A projectile is fired from ground level towards the wall of a building 640 ft. distant. The angle of projection, θ, with the horizontal ground is not given, but the initial velocity has a <u>vertical component</u> of 96 ft/sec upwards. Five seconds after it is fired, <u>the projectile strikes the wall</u>.

(a) What is the maximum height of the projectile above ground?
(b) How far above ground does the projectile hit the wall?
(c) Find the velocity with which the projectile hits the wall. Remember that velocity is a <u>vector</u>!

* *

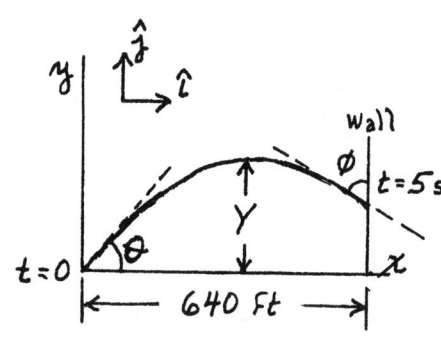

Projectile motion (air neglected) has constant downward acceleration "g". Use horizontal (x) and vertical (y) axes to locate projectile, $v_x = dx/dt$, $v_y = dy/dt$, and
$$a_x = \frac{dv_x}{dt} = 0 \quad a_y = \frac{dv_y}{dt} = \text{constant.}$$

Integration gives (1) $v_x = v_{xo}$ (3) $v_y = v_{yo} + a_y t$
                (2) $x = x_o + v_x t$ (4) $y = y_o + v_{yo} t + \frac{1}{2} a_y t^2$

Eliminating t from (3),(4) gives $v_y^2 = v_{yo}^2 + 2a_y(y-y_o)$ (5)
     " a_y "   "   " $\frac{1}{2}(v_y + v_{yo}) = (y-y_o)/t$ = Avg y-velocity

(a) Let t_r = time to rise to peak, where $v_y = 0$, $y = Y$ (maximum)
In Eq. (3): $0 = v_{yo} + a_y t_r = +96$ ft/s $+(-32$ ft/s$^2) t_r \Rightarrow t_r = 3$ s.
In (4): $Y = 0 + 96(3) + \frac{1}{2}(-32) 3^2 = 144$ ft

OR in Eq (5): $0 = (96)^2 + 2(-32) Y \Rightarrow Y = 144$ ft

(b) In (4), at $t = 5$, $y = 0 + 96 \cdot 5 + \frac{1}{2}(-32) 5^2 = 80$ ft, above the ground.

(c) Must find v_x, v_y for $t = 5$s
From (2) $640 = v_x \cdot 5 \Rightarrow v_x = v_{xo} = 128$ ft/s
From (3) $v_y = 96 + (-32) \cdot 5 = -64$ ft/s.

∴ $\vec{v}(t=5) = 128 \hat{i} - 64 \hat{j}$ ft/s
or $v = \sqrt{(128)^2 + (64)^2} = 143$ ft/s at an angle $\phi = \tan^{-1}\left(\frac{128}{64}\right) = 63.4°$ with the wall.

4-42

A student shoots a projectile of diameter 1.2 cm and mass 45 grams horizontally from a spring-loaded gun in the Physics 211 laboratory. The projectile falls through a vertical distance of 80 cm while traveling a horizontal distance of 105 cm, landing on a horizontal surface.
a) How long was the projectile in flight?
b) What was the initial velocity of the projectile?
c) What was the vertical component of the velocity just before it struck the surface?
d) What was the horizontal component of the velocity just before it struck the surface?

****

a) Let y be up, Let x be in direction of initial v

Let $x=0$, $y=0$, $t=0$ at firing time

$$y = -\tfrac{1}{2}(9.81)t^2$$

for $y = -0.80$ m, $t^2 = \dfrac{1.60}{9.8}$

time in air $= \sqrt{\dfrac{1.60}{9.8}} = 0.404$ s

b) $x = v_0 t$

$t = 0.404$ s $x = 1.05$ m

$\Rightarrow v_0 = \dfrac{1.05}{0.404} = 2.60$ m/s

c) $v_y = -9.8\,t = -9.8(0.404) = -3.96$ m/s

d) $v_x = v_0 = 2.60$ m/s

4-43

A motorcyclist drives up and off a cement incline which has an inclination angle of 32.2° and lands 32.2 meters, along the ground, from the point where he left the incline. If the motorcycle moved 32.2 feet along the incline before leaving the incline, at what speed was the driver maintaining along the incline assuming a constant velocity for the entire 32.2 feet?

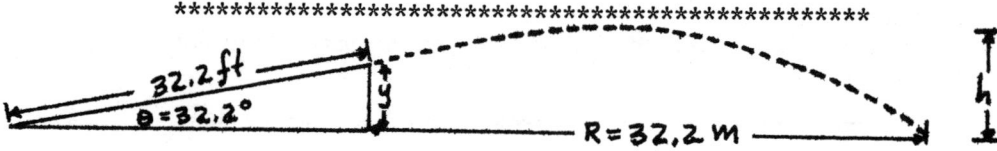

$$y = 32.2 \text{ ft} \sin 32.2° = 17.159 \text{ ft}$$

$$R = \text{range} = 32.2 \text{ m} \left(\frac{3.281 \text{ ft}}{1 \text{ m}}\right) = 105.648 \text{ ft}$$

By knowing that the vertical velocity at the top of a projectile's flight is $v_{y_{top}} = v_{oy} - g t_{top} = 0$ (1)

then the time required to move a vertical upward distance of $h - y$ is

$$(2) \quad t_{top} = \frac{v_{oy}}{g} = \frac{v_o \sin\theta}{g}$$

The time of the motorcycle's downward flight a distance y is found from

$$(3) \quad y = v_{oy} t + \tfrac{1}{2} g t^2 = (v_o \sin\theta) t + \tfrac{1}{2} g t^2$$

or from the quadratic equation,

$$(4) \quad t = \frac{-v_o \sin\theta + \sqrt{v_o^2 \sin^2\theta + 2gy}}{g}$$

Hence, the time of the entire trip, from (2) and (4), is

$$(5) \quad t_{trip} = 2 t_{top} + t$$

$$= \frac{2 v_o \sin\theta}{g} - \frac{v_o \sin\theta}{g} + \frac{\sqrt{v_o^2 \sin^2\theta + 2gy}}{g}$$

$$= \frac{v_o \sin\theta + \sqrt{v_o^2 \sin^2\theta + 2gy}}{g}$$

The range, R, is given by the expression

$$(6) \quad R = v_{ox} t_{trip} = (v_o \cos\theta) t_{trip}$$

Substitution of expression (5) into (6) leads to

(7) $R = v_0 \cos\theta \left[\dfrac{v_0 \sin\theta + \sqrt{v_0^2 \sin^2\theta + 2gy}}{g} \right]$

Now we have to solve expression (7) for the motorcycle's velocity off the incline.

$R = \dfrac{v_0^2 \sin\theta \cos\theta}{g} + \dfrac{v_0 \cos\theta \sqrt{v_0^2 \sin^2\theta + 2gy}}{g}$

$\left(R - \dfrac{v_0^2 \sin\theta \cos\theta}{g} \right)^2 = \left(\dfrac{v_0 \cos\theta \sqrt{v_0^2 \sin^2\theta + 2gy}}{g} \right)^2$

$R^2 - \dfrac{2R v_0^2 \sin\theta \cos\theta}{g} = \dfrac{(v_0^2 \cos^2\theta)(2gy)}{g^2}$

$gR^2 - 2R v_0^2 \sin\theta \cos\theta = 2y v_0^2 \cos^2\theta$

$v_0^2 (2R \sin\theta \cos\theta + 2y \cos^2\theta) = gR^2$

(8) $v_0 = R \sqrt{\dfrac{g}{2}} \sqrt{\dfrac{1}{R \sin\theta \cos\theta + y \cos^2\theta}}$

We are now in a position to substitute the values of R, g, y and θ into expression (8) to arrive at v_0:

$v_0 = 105.648 \text{ ft} \sqrt{\dfrac{32.2 \text{ ft/sec}^2}{2}} \cdot \sqrt{\dfrac{1}{105.648 \text{ ft} \sin 32.2° \cos 32.2° + 17.159 \text{ ft} \cos^2 32.2°}}$

$v_0 = 54.761 \text{ ft/sec}$

4-44

A baseball hit out of a ballpark left the bat with a velocity of 36 m/s, an angle of 50° above the horizontal, and a height of 1.2 m above the ground.

a) How high above the ground was the ball when it passed over the wall, a distance of 122 m from home plate?

b) How far did the ball travel horizontally before striking the ground?

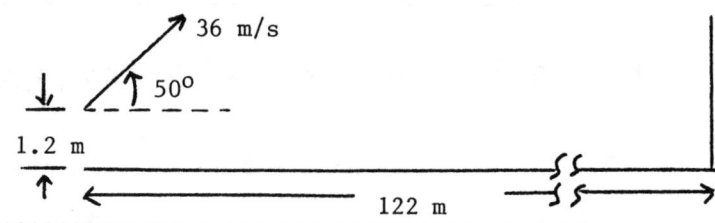

****

a) Careful here; since $y_0 \neq 0$ you cannot use the range equation which assumes $x_0 = y_0 = 0$.

$x_0 = 0$ $v_{0x} = v_0 \cos 50° = 36 \text{ m/s} \times 0.643 = 23.1 \text{ m/s}$

$y_0 = +1.2 \text{ m}$ $v_{0y} = v_0 \sin 50° = 36 \text{ m/s} \times 0.766 = 27.6 \text{ m/s}$

$x = x_0 + v_{0x} t$ or $122 \text{ m} = 0 + 23.1 \text{ m/s} \times t$

$t = \dfrac{122 \text{ m}}{23.1 \text{ m/s}} = 5.28 \text{ s}.$

$y = y_0 + v_{0y} t + \tfrac{1}{2} g t^2$

$y = 1.2 \text{ m} + 27.6 \text{ m/s} \times 5.28 \text{ s} + \tfrac{1}{2}(-9.8 \text{ m/s}^2)(5.28 \text{ s})^2$

$y = \underline{10.3 \text{ m}}$ ∴ ball is 10.3 m above ground at wall.

b) When the ball reaches the ground we have $y = 0$.

$y = y_0 + v_{0y} t + \tfrac{1}{2} g t^2$

$0 = 1.2 \text{ m} + 27.6 \text{ m/s} \, t + \tfrac{1}{2}(-9.8 \text{ m/s}^2) t^2$

$$4.9 \text{ m/s}^2 \, t^2 - 27.6 \text{ m/s} \, t - 1.2 \text{ m} = 0$$

$$t = \frac{27.6 \text{ m/s} \pm \sqrt{(27.6 \text{ m/s})^2 + 4 \times 4.9 \text{ m/s}^2 \times 1.2 \text{ m}}}{2 \times 4.9 \text{ m/s}^2}$$

$$t = \frac{27.6 \text{ m/s} \pm 28.0 \text{ m/s}}{9.8 \text{ m/s}^2} = +5.68 \text{ s}$$

(only the positive root is appropriate here)

$$x = x_0 + v_{0x} t = 0 + 23.1 \text{ m/s} \times 5.68 \text{ s} = \underline{131.2 \text{ m}}$$

4-45

A rock is tossed from a bridge in the upward direction at speed 3.0 m/sec, and it lands in the water below the bridge 4.0 sec later. Calculate the height of the launch point relative to the water, and the speed with which the rock hits the water.

****

altitude vs. time ($t_1 = 0$)

$$y_2 = y_1 + v_1 t_2 + \tfrac{1}{2} a_y t_2^2$$

Set $y_2 = 0$, $v_1 = 3$, $t_2 = 4$, $a_y = -g$

($g = 9.8 \text{ m/sec}^2$)

Then $y_1 = \underline{66.4 \text{ m}}$

also, $v_2 = v_1 + a_y t_2 = 3 - 9.8 \times 4 = -36.2 \text{ m/sec}$,

so impact speed is $\underline{36.2 \text{ m/sec}}$.

4-46

As shown in the figure below, a projectile is fired with a muzzle velocity of 50.0 m/s at an angle of 36.87° above the horizontal. At a distance of 80.0 meters to the right is a vertical cliff 20.0 meters high. Determine where the projectile first collides with the ground, cliff wall, or cliff top. Give both x and y coordinates. (Use 9.8 m/s² ≅ 10 m/s² as the acceleration due to gravity.)

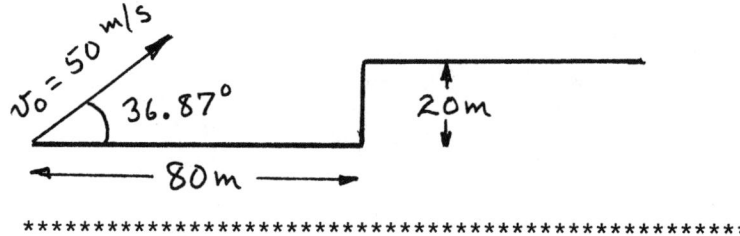

****

The Master Equation for motion with constant acceleration, $s = s_0 + v_0 t + \frac{1}{2} a t^2$, can be specialized for x motion and y motion independently. Choosing a coordinate system with the origin at the projectile's starting point, the initial conditions are

x-motion	y-motion
$x_0 = 0$	$y_0 = 0$
$v_{0x} = v_0 \cos \theta = 40$ m/s	$v_{0y} = v_0 \sin \theta = 30$ m/s
$a_x = 0$	$a_y = -10$ m/s²

The master equation becomes:

$x = x_0 + v_{0x} t + \frac{1}{2} a_x t^2$	$y = y_0 + v_{0y} t + \frac{1}{2} a_y t^2$
$x = (40 \text{ m/s}) t$	$y = (30 \text{ m/s}) t - (5 \text{ m/s}^2) t^2$

We can determine the fate of the projectile by tracing its path at a few instants of time using the x and y equations:

t	0s	1s	2s	3s	4s	5s
x	0 m	40	80	120	160	200
y	0 m	25	40	45	40	25

The projectile clearly clears the cliff at $t = 2$ sec. Since the projectile hits atop the cliff, we ask when does $y = 20$ m?

Using $y = 30t - 5t^2 = 20$ and the quadratic equation, we have

$$t = \frac{30 \pm \sqrt{30^2 - 4(5)(20)}}{10}$$

$$= \frac{30 \pm 22.36}{10} = 0.764 \text{ sec and } 5.236 \text{ s}$$

The first root (0.764 s) is when the projectile passed the 20 m mark on the way up (see t, x, y table). Thus the second root (5.236 sec) is the desired time and

$$x = 40t = \underline{\underline{209 \text{ m}}}, \quad y = \underline{\underline{20 \text{ m}}}$$

CIRCULAR MOTION

4-47

An artificial satellite circles the earth in a circular orbit of radius 6.6×10^6 m. If it completes one orbit in 89 minutes, estimate the value of g at this altitude.

****

The orbital speed of the satellite v is

$$v = \frac{2\pi (6.6 \times 10^6)}{89 \times 60} \quad \text{m/sec}.$$

The centripetal acceleration is then

$$\frac{(2\pi)^2 (6.6 \times 10^6)^2}{(89 \times 60)^2} \times \frac{1}{(6.6 \times 10^6)} \quad \text{m/sec}^2,$$

which is the value of g at this altitude and it turns out to be approximately 9.14 m/sec².

4-48

A 2 kg mass is attached to a vertical rod by two ropes as shown in the diagram below. The ropes are under tension when the rod is rotated uniformly about its axis. If the speed of the mass is 3 m/s as it moves in circular motion, find the tension in each rope.

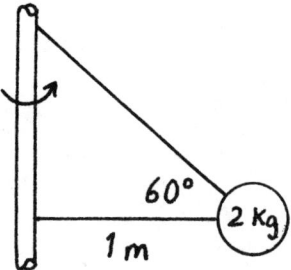

****

force diagram

$\Sigma F_y = 0$

$P \sin 60° - mg = 0$

$P = \dfrac{mg}{\sin 60°} = 22.6 N$

$\Sigma F_x = ma_c = m\dfrac{v^2}{r}$ r = radius = 1 m

$T + P \cos 60° = m\dfrac{v^2}{r}$

$T = \dfrac{mv^2}{r} - P \cos 60° = (2)\dfrac{3^2}{1} - (22.6) \cos 60°$

$T = 6.7 N$

4-49

A 1500 kg car is racing on a horizontal circular track of circumference 628 meters. The track surface is level and has a coefficient of friction of 0.70 with the car's tires. The car makes one complete circuit of the track every 40 seconds at constant speed.
a) Draw a free body diagram of the car on the track.
b) Determine the magnitude of the resultant of all the forces acting on the car. State specifically its direction.
c) What is the maximum speed the car can have without slipping?

****

a)

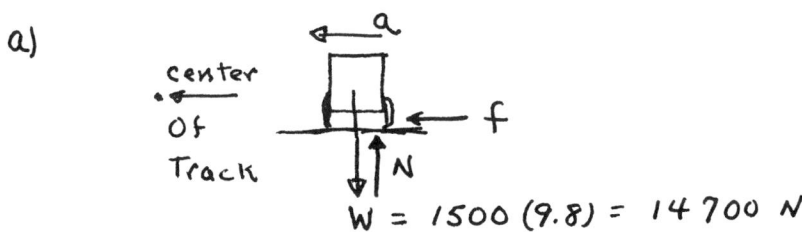

$W = 1500(9.8) = 14700$ N

b) The resultant force is in the direction of the acceleration or toward the center of the circular track. The magnitude is

$$F = ma$$

and

$$a = \frac{v^2}{r}, \quad v = \frac{628}{40} = 15.7 \frac{m}{s}$$

$$r = \frac{628}{2\pi} = 100 \text{ m}$$

Thus: $a = \frac{(15.7)^2}{100} = 2.46 \text{ m/s}^2$

$F = 2.46(1500) = 3697$ N Toward Center

c) The source of this force is the static friction between the tires and the track

$$f_{max} = \mu N$$

$\Sigma F_{vert} = 0 ; \Rightarrow N = W = mg$

$\Rightarrow f_{max} = \mu mg = \frac{m(v_{max})^2}{r}$

$\Rightarrow v_{max} = \sqrt{\mu g r} = \sqrt{0.7(9.8)(100)}$

$= 26.2$ m/s

4-50

(a) A stone of mass 1 kg is attached to one end of a string one meter long. It is whirled in a horizontal circle on a frictionless table top, with a velocity of 22.5 meters per second. The other end of the string is kept fixed. What is the tension in the string?

(b) The net force acting on the moon is the sum of two forces, the centripetal force and the gravitational force due to the earth. True or False? Explain your answers.

****

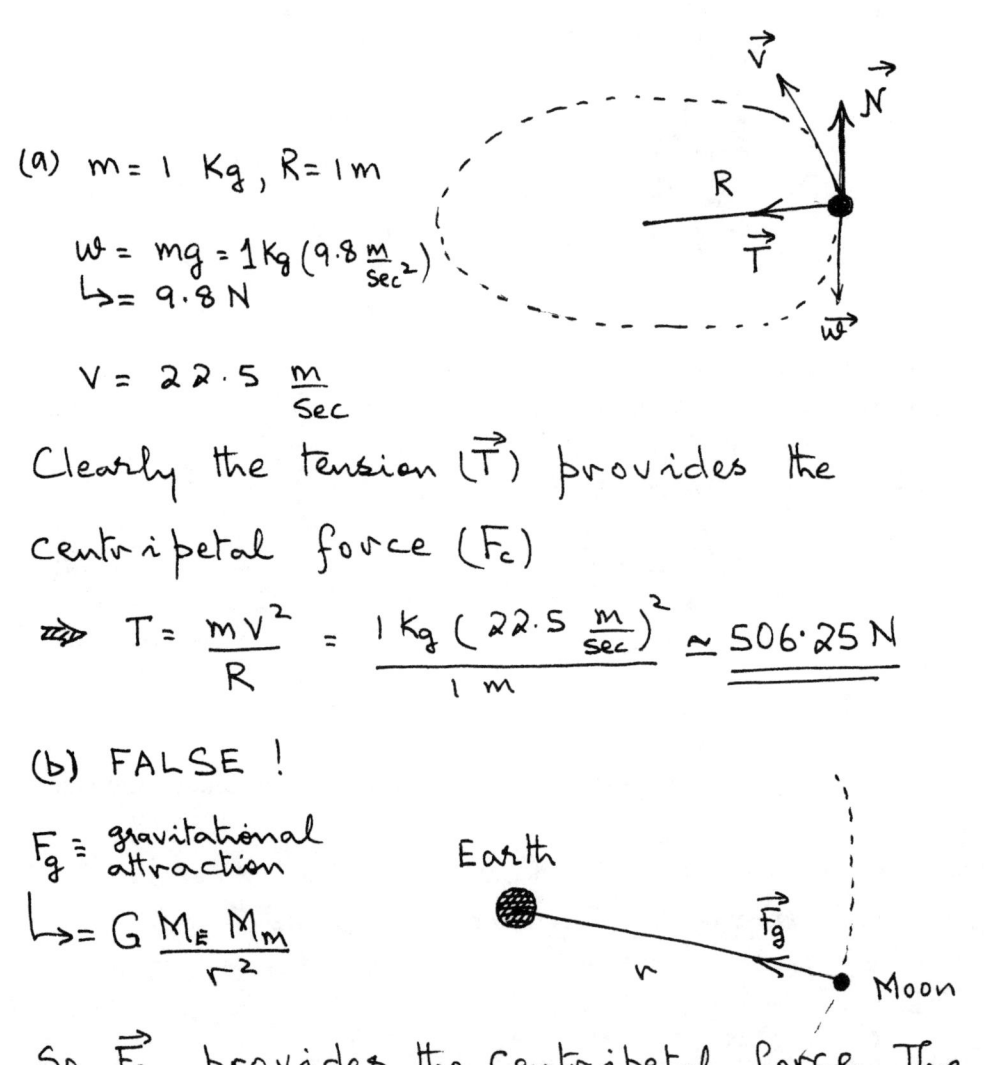

(a) $m = 1 \text{ Kg}, R = 1 \text{ m}$

$W = mg = 1 \text{Kg}(9.8 \frac{m}{sec^2})$
$\hookrightarrow = 9.8 \text{ N}$

$V = 22.5 \frac{m}{sec}$

Clearly the tension ($\vec{T}$) provides the centripetal force (F_c)

$\Rightarrow T = \frac{mv^2}{R} = \frac{1 \text{Kg}(22.5 \frac{m}{sec})^2}{1 \text{ m}} \approx \underline{\underline{506.25 \text{ N}}}$

(b) FALSE!

$F_g =$ gravitational attraction
$\hookrightarrow = G \frac{M_E M_m}{r^2}$

So $\vec{F_g}$ provides the centripetal force. The two forces are therefore one and the same!

4-51

A plumb bob does not hang along a line through the center of the earth. Instead, because of the rotation of the earth, it is displaced outward (toward the south in the northern hemisphere) through a certain angle. (a) Find the angle as a function of latitude. (b) Find the latitude of maximum angular displacement. (c) Find the angular displacement for a latitude of 45°.

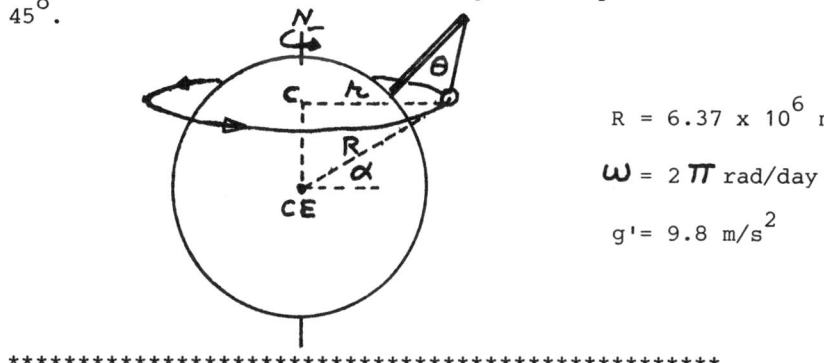

$R = 6.37 \times 10^6$ m

$\omega = 2\pi$ rad/day

$g' = 9.8$ m/s^2

***

(a) g' IS THE EFFECTIVE GRAVITATIONAL FIELD MEASURED IN THE LABORATORY. T IS THE TENSION IN THE STRING. F IS THE FORCE OF GRAVITY WHICH IS DIRECTED TOWARD THE CENTER OF THE EARTH (ASSUMED TO BE A UNIFORM SPHERE).

THE VECTOR SUM OF $\vec{F}$ AND $\vec{T}$ EQUALS $\vec{F_c}$, THE CENTRIPETAL FORCE, WHICH MUST POINT TOWARD THE CENTER OF THE CIRCLE OF MOTION C, NOT TOWARD THE CENTER OF THE EARTH CE.

$T = mg'$, $F_c = mr\omega^2 = m(R\cos\alpha)\omega^2$

$$\frac{\sin\theta}{F_c} = \frac{\sin\alpha}{T}$$

$$\sin\theta = \frac{F_c}{T}\sin\alpha$$

$$\sin\theta = \frac{R\omega^2}{g'}\cos\alpha \sin\alpha = \frac{R\omega^2}{2g'}\sin 2\alpha$$

(b) $\alpha = 45°$ FOR MAX θ. (c) $\sin\theta = \frac{(6.37 \times 10^6)}{2(9.8)}\left[\frac{2\pi}{24(3600)}\right]^2$

$\theta_{MAX} = 0.098°$

4-52

A 1500-kg car initially going North turns West by going around the circular arc, APB, with uniform speed. The arc length, APB, is 200 m, and the car completes the turn in 10 seconds. In your answers, leave any factors of π indicated.

(a) Find the acceleration of the car when it is at point P, two-thirds of the way around the curve. Write the answers using unit vectors $\hat{i}$ (East) and $\hat{j}$ (North).
(b) Find the AVERAGE velocity of the car during the ten seconds it took to complete the turn.
(c) Find the AVERAGE acceleration of the car during the ten seconds.

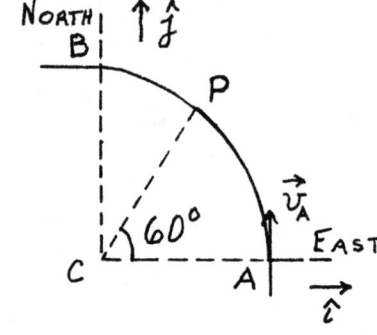

* *

(a) The car, with constant speed $v = 200 \text{ m}/10 \text{ s} = 20$ m/s, has a centripetal acceleration $a_c = v^2/R$ directed from car to center of circle.

$\widehat{APB} = 200 \text{ m} = \frac{1}{4}(\text{circumference}) = \frac{1}{4}(2\pi R)$ so
$R = 400/\pi$ m and
$a_c = v^2/R = 20^2/(400/\pi) = \pi \text{ m/s}^2$, 60° S of W

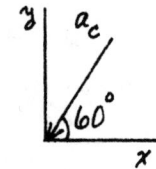

$(a_c)_x = -\pi \cos 60° = -0.500\pi$
$(a_c)_y = -\pi \sin 60° = -0.866\pi$ or
$\vec{a}_c = -\pi(0.500\hat{i} + 0.866\hat{j})$ m/s²

(b) Let $\vec{\Delta R}$ = displacement from A to B = $-R\hat{i} + R\hat{j}$
$= \frac{400}{\pi}(-\hat{i} + \hat{j})$
By definition, Avg $\vec{v} = \vec{\Delta R}/\Delta t = \frac{40}{\pi}(-\hat{i} + \hat{j})$
$= \frac{40}{\pi}\sqrt{2} = 18.0$ m/s NW

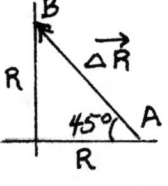

(c) By definition, Avg $\vec{a} = \vec{\Delta v}/\Delta t = (\vec{v}_B - \vec{v}_A)/\Delta t$
$= \frac{-20\hat{i} - (20\hat{j})}{10 \text{ s}} = -2(\hat{i} + \hat{j})$ m/s²
or $2\sqrt{2}$ m/s² SW

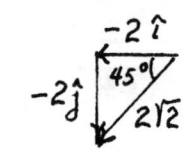

4-53

A boy is riding on a merry-go-round which has a radius of 5.0 m and makes 10 revolutions per minute. His friend is standing on the ground nearby watching. The boy is carrying a ball. When the merry-go-round comes around and he is directly facing his friend (along the tangent to the circle), he throws the ball straight up in the air (in his own reference frame) with a velocity of 7.0 m/s. The ball eventually falls right into the hands of his friend. Assume that the height of the boy's hand, off the ground, when he throws the ball, is the same as that of his friend when he catches it.

(a) What is the total acceleration of the ball <u>before</u> the boy throws it?
(b) What is the total acceleration of the ball <u>after</u> he throws it?
(c) What is the velocity of the ball, in the reference frame of the friend, just after the boy throws it?
(d) How far away from the center of the merry-go-round is the friend standing?

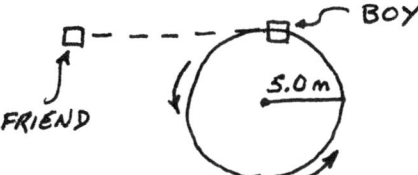

****

(a) $a = \dfrac{v^2}{R}$, $v = \dfrac{2\pi R}{T}$, $R = 5.0$ m, $T = 0.1$ min $= 6.0$ s

$\Rightarrow$ $\boxed{\vec{a} = 5.5 \text{ m/s}^2 \text{ DIRECTED TOWARD CENTER OF MERRY-GO-ROUND}}$

(b) $\boxed{a = g = 9.8 \text{ m/s}^2 \text{ DIRECTED DOWN}}$

(c) $v_{HORIZONTAL} = \dfrac{2\pi R}{T} = 5.2$ m/s

$v_{VERTICAL} = 7.0$ m/s

$\Rightarrow$ $\boxed{v = 8.7 \text{ m/s DIRECTED UPWARD AT } 53° \text{ ANGLE!}}$

(d) EQUNS OF MOTION, AFTER BALL IS THROWN, ARE:

$\Delta y = v_y^0 t - \tfrac{1}{2} g t^2$, $\Delta x = v_x^0 t$, $v_x^0 = 5.2$ m/s

BALL IS CAUGHT AT SAME HEIGHT AS THROWN $\Rightarrow \Delta y = 0$.

$v_y^0 = 7.0$ m/s $\Rightarrow t = 1.4$ s $\Rightarrow \Delta x = 7.5$ m

$\boxed{\text{DIST. FROM CENTER} = \sqrt{R^2 + (\Delta x)^2} = 9.0 \text{ m}}$

4-54

A particle undergoes uniform circular motion with a period of 10 seconds and a radius of 2 meters, starting at a point whose x coordinate is 2 and whose y coordinate is zero and rotating counter-clockwise.

Find:
(a) the average velocity vector in the time interval from t = 2.5 to t = 5 seconds;
(b) the instantaneous velocity vectors at both t = 2.5 and t = 5 seconds;
(c) the average acceleration vector during the time interval from t = 2.5 to t = 5 seconds; and
(d) the instantaneous acceleration vector when the time is 2.5 seconds.

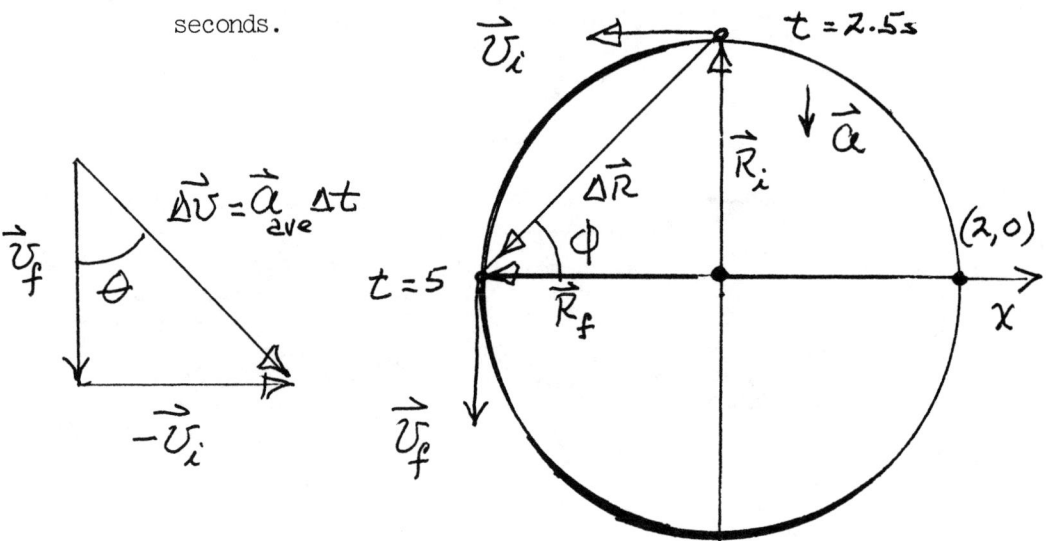

****

a) $\vec{v}_{ave} = \dfrac{\Delta \vec{R}}{\Delta t} = \dfrac{\vec{R}_f - \vec{R}_i}{\Delta t} = \dfrac{-2\vec{i} - 2\vec{j}\, m.}{2.5\,s} = -0.8(\vec{i}+\vec{j})$

The magnitude of $\vec{v}_{ave}$ and the angle ϕ are

$|\vec{v}_{ave}| = \sqrt{2(0.8)^2} = 1.13\, m/s\,;\ \tan\phi = 1\,;\ \phi = 45°$

Hence $\vec{v}_{ave}$ makes an angle of $135°$ with $+x$ axis.

b) $\vec{v}_i = -(speed)\vec{i} = -1.26\,\vec{i}\,;\ \vec{v}_f = -(speed)\vec{j} = -1.26\,\vec{j}$

Where speed $= 2\pi R/T = \dfrac{(6.28)(2m.)}{10s} = 1.26$ m/s

c) $\vec{a}_{ave} = \dfrac{\Delta \vec{v}}{\Delta t} = \dfrac{\vec{v}_f - \vec{v}_i}{\Delta t} = \dfrac{-1.26\vec{j} + 1.26\vec{i}}{2.5s}$ m/s

$\vec{a}_{ave} = +(0.5)(\vec{i} - \vec{j})$ m/s^2. Note that the average value calculated here is obtained by subtracting two instantaneous values from each other.

The magnitude of $\vec{a}_{ave}$ and the angle θ are

$|\vec{a}_{ave}| = \sqrt{2(0.5)^2} = 0.7$ m/s^2 ; $\tan \theta = 1$; $\theta = 45°$

Hence $\vec{a}_{ave}$ makes an angle of 315° with the +x axis.

d) $\vec{a} = -\dfrac{v^2}{R}\vec{j} = -\dfrac{(1.26)^2}{2}\vec{j} = -0.8\vec{j}$ m/s^2

Note that v used above is an instantaneous velocity.

4-55

A car moves on a circular track banked at angle $\theta = 30°$ and having radius $R = 45$ m. The test driver finds that the maximum value of constant speed he can drive the car along the track without sliding is 35 m/s.

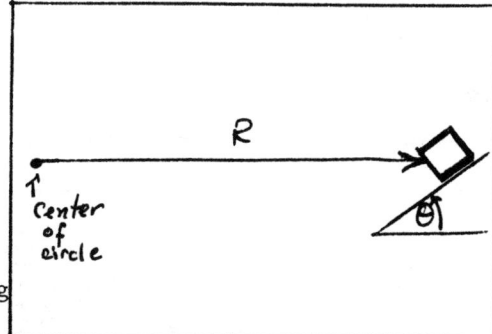

(a) Draw a free-body diagram indicating the forces acting on the car.

(b) Using the free body diagram as a guide, set up appropriate coordinate axes and find the coefficient of static friction between car and road.

(c) Find the *period* of the motion of the car on the circular track.

(d) Find the *frequency* for this motion.

****

(a)

$\vec{f}_s$ is the static friction force, $\vec{N}$ the normal force, and $\vec{W}$ the weight of the car.

(b)

z-equation
$$N\cos\theta - f\sin\theta - mg = 0 \quad (1)$$
$$a_z = 0$$

r-equation
$$N\sin\theta + f\cos\theta = m\frac{v^2}{R} \quad (2)$$

Using $f_s = \mu_s N$ in Eqs. (1) and (2) we have a solution in the two unknowns μ_s and N:

Using Eq (1), $\quad N = \dfrac{mg}{\cos\theta - \mu_s \sin\theta} \quad (3)$

Substitution of ③ into ② gives

$$\left(\frac{mg}{\cos\theta - \mu_s \sin\theta}\right)\sin\theta + \mu\left(\frac{mg}{\cos\theta - \mu_s \sin\theta}\right)\cos\theta = m\frac{v^2}{R}$$

Substitution of the given values for θ, v, and R gives $\mu_s = 0.85$.

(c) The period T is the time required for the car to make one complete circuit of the track.

$$T = \frac{2\pi R}{v} = \frac{2\pi \times 45}{35} = 8.1 \text{ s}$$

(d) The frequency f, the number of cycles or the number of circuits of the track per unit time, is the reciprocal of the period T.

$$f = \frac{1}{T} = \frac{1}{8.1 \text{ s}} = 0.12 \text{ s}^{-1}$$

4-56

An object being swung on a string in a horizontal circle of a radius of 2.3m at a height of 6.2m above the ground makes two revolutions in 0.48s. How far will the object travel if the string is released?

****

Describing problem:

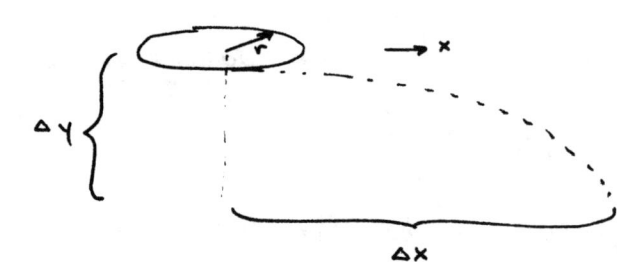

$r = 2.3\,m$ $T = \frac{.48}{2} = .24\,s$ $v_{oy} = 0$
$\Delta y = -6.2\,m$ $a_y = -g = -9.8\,m/s^2$

Setting up eqs.

$$v_x = \frac{C}{T} = \frac{2\pi r}{T} = \frac{\Delta x}{\Delta t} \Rightarrow \Delta x = 2\pi r \frac{\Delta t}{T}$$

and

$$\Delta y = \overset{0}{v_{oy}} \Delta t + \frac{1}{2}\overset{-g}{a_y} \Delta t^2$$

$$\Rightarrow \Delta t^2 = -\frac{2\Delta y}{g} \Rightarrow \Delta t = \sqrt{\frac{-2\Delta y}{g}}$$

So

$$\Delta x = \frac{2\pi r}{T}\sqrt{\frac{-2\Delta y}{g}}$$

Substituting numbers:

$$\Delta x = \frac{2\pi(2.3)}{.24}\sqrt{\frac{-2(-6.2)}{9.8}} = \underline{\underline{68\,m}}$$

4-57

The body at point C has a mass of 20 kg and moves in a horizontal circular path (as shown by the dashed circle) about the vertical axis AB, with a constant speed of 8 meters per second.
(a) What is the resultant vertical force acting on the body?
(b) What is the resultant horizontal force acting on the body?
(c) What is the tension in string AC?
(d) What is the tension in string BC?
(e) What is the acceleration of the body?

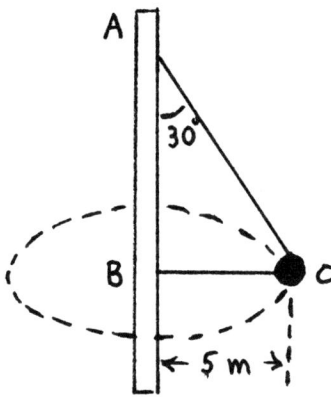

****

Free body diagram of the 20 kg body:

(a) Since the body is moving in a horizontal plane, $\underline{\underline{\Sigma F_y = 0}}$

(b) $\Sigma F_x = \dfrac{mv^2}{r} = \dfrac{(20)(8)^2}{5} = 256\ N \leftarrow$

(c) $+\uparrow\ \Sigma F_y = 0 = T_1 \cos 30 - 20(9.8),\quad T_1 = 226\ N\ \underline{60}$

(d) $\xleftarrow{+}\ \Sigma F_x = T_2 + 226 \sin 30 = 256$ where

$T_1 = 226$ and $\Sigma F_x = 256$ as per part (b) $\therefore T_2 = 143\ N \leftarrow$

(e) $a = \dfrac{v^2}{R} = \dfrac{8^2}{5} = 12.8\ m/s^2 \leftarrow$

4-58

A car on a flat (unbanked) circular track accelerates uniformly from rest with a forward (tangential) acceleration of 0.80 m/s². What must be the coefficient of friction between the track and car in order for the car to get half-way around the track before it "spins-out"?

****

Describing problem:

$\theta_{1/2} = \pi$

$a_t = 0.8 \, m/s$

$v_0 = 0$

"Spin out"

Setting up eqs.

$$\left. \begin{array}{l} a_c = \dfrac{v^2}{R} \\ v^2 = \cancel{v_0^2} + 2a_t s \end{array} \right\} \Rightarrow a_c = \dfrac{2 a_t s}{R} = \dfrac{2 a_t R\theta}{R} = 2 a_t \theta$$

Then

$$a = \sqrt{a_t^2 + a_c^2} = \sqrt{a_t^2 + 4 a_t^2 \theta^2} = a_t \sqrt{1 + 4\theta^2}$$

Now

$$\left. \begin{array}{l} f = \mu F_N = \mu mg \\ \text{From Newton's 2nd law} \\ f = ma \end{array} \right\} \Rightarrow a = \mu g$$

$$\Rightarrow \mu g = a_t \sqrt{1 + 4\theta^2}$$

$$\Rightarrow \mu = \dfrac{a_t}{g} \sqrt{1 + 4\theta^2}$$

Substituting numbers:

$$\mu = \dfrac{0.8}{9.8} \sqrt{1 + 4\pi^2} = \underline{\underline{0.52}}$$

4-59

A rider in a Ferris wheel experiences a force against his seat equal to 150% of his weight when he moves through the bottom of the circle. The radius of the wheel is 24 meters. What is the tangential velocity of the rider?

****

Construct a free-body diagram for the rider.

↑ N
↓ W

$F = ma$

$N - W = m\left(\dfrac{v^2}{r}\right) = \dfrac{W}{g} \dfrac{v^2}{r}$

$1.5W - W = \dfrac{Wv^2}{gr}$

$0.5 = \dfrac{v^2}{rg}$

$v = \sqrt{0.5\, rg} = \sqrt{0.5(24m)(9.8\, m/sec^2)}$

$\underline{v = 10.8\ m/sec}$

NEWTON'S LAW OF GRAVITATION

4-60

Data from astronomy suggest that our solar system is in an essentially circular orbit around our galaxy with a period of 8×10^{15} seconds and a radius of 3×10^{20} meters. Find the mass at the center of this orbit which would account for this uniform circular motion. Express your answer as a multiple of the sun's mass. The mass of the sun (a typical star) is about 2×10^{30} kg. Hint: $4\pi^2/G = 6 \times 10^{11}$ kg sec^2 m^{-3} where G is the constant in Newton's law of gravitation.

****

$$\frac{GMM'}{R^2} = \frac{Mv^2}{R} \Rightarrow \frac{GM'}{R^2} = \frac{v^2}{R} = \omega^2 R$$

$$\Rightarrow GM' = \omega^2 R^3 = \left(\frac{2\pi}{T}\right)^2 R^3 = \frac{4\pi^2 R^3}{T^2}$$

$$\Rightarrow M' = \frac{4\pi^2}{G} \frac{R^3}{T^2} = 6(10)^{11} \frac{kg\ sec^2}{m^3} \frac{(3 \times 10^{20} m)^3}{(8 \times 10^{15} sec)^2}$$

$$M' = \frac{6(27)}{64} (10)^{11+60-30}\ kg = 2.5 \times 10^{41}\ kg$$

$$M' = 2.5 \times 10^{41}\ kg \left[\frac{1\ sun}{2 \times 10^{30}\ kg}\right]$$

$$\boxed{M' = 1 \times 10^{11}\ suns}$$

4-61

Two spherical masses M and 2M are separated by a distance R as measured from their centers.

a) Calculate the position of a point on a line between the two centers where the net gravitational force on a test particle is zero.
b) Calculate the gravitational potential energy of a mass 3M placed midway between the two original masses.

****

a) $$\frac{2MmG}{x^2} = \frac{MmG}{(R-x)^2}$$

$$\frac{2}{x^2} = \frac{1}{(R-x)^2}$$

$$\frac{x^2}{2} = R^2 - 2Rx + x^2$$

$$x^2 = 2R^2 - 4Rx + 2x^2$$

$$x^2 - 4Rx + 2R^2 = 0$$

$$x = \frac{4R \pm \sqrt{16R^2 - 4(2R^2)}}{2} = \frac{4R \pm \sqrt{8R^2}}{2}$$

$$= \frac{4R \pm 2\sqrt{2}R}{2} = (2 \pm \sqrt{2})R$$

$x_- = (2 - 1.41)R \approx 0.6R$
$x_+ = (2 + 1.41)R \approx 3.4R$ Not $< R$.

b) $$U = -\frac{(3M)(2M)G}{R/2} - \frac{(3M)(M)G}{R/2}$$

$$= -\frac{6M^2G + 3M^2G}{R/2} = -\frac{18M^2G}{R}$$

4-62

Two spherical shells of mass M_1 and M_2 and of radius R_1 and R_2 are situated as shown; their center-to-center distance is called D. Find the force on the point mass m, which is located a distance r from the center of shell 1.

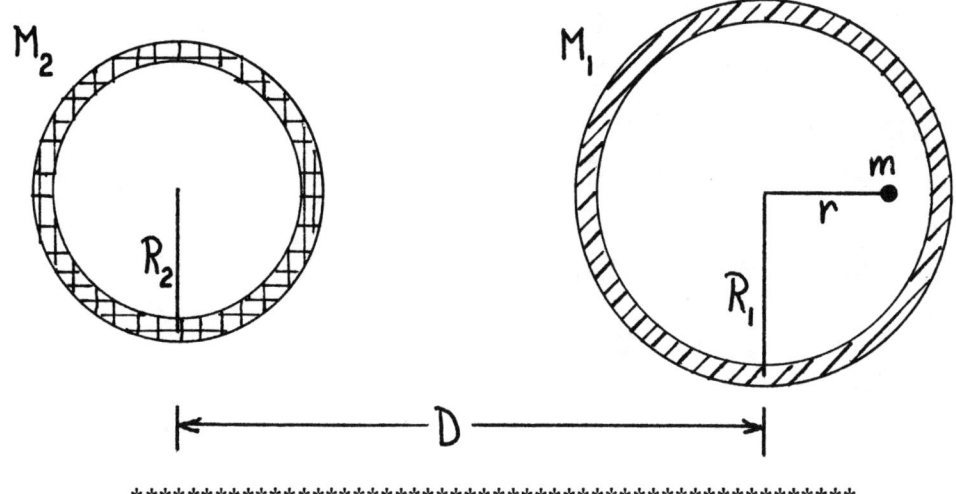

***

Shell #1 has <u>no</u> effect on m; such shells create no gravitational forces on particles inside.

Shell #2 acts (on m, which is exterior to it) as tho' its mass, M_2, were located at its center. So the equivalent problem is:

$$M_2 \longleftarrow D+r \longrightarrow m$$

and the force on m is $\dfrac{G M_2 m}{(D+r)^2}$.

4-63

A mass M in the form of a thin uniform semicircle exerts a gravitational force on a small mass located at the center. (a) Find the magnitude and direction of the force. (b) Find the center of mass of M. (c) Find the center of gravity of M under the given conditions.

(a) $dF = \dfrac{Gm\,dM}{R^2}$

$\lambda = \dfrac{dM}{d\ell} = \text{const.} = \dfrac{M}{\ell} = \dfrac{M}{\pi R}$

$dM = \lambda\,d\ell = \lambda R\,d\theta$

$dF_x = dF\cos\theta = \dfrac{Gm\lambda R\cos\theta\,d\theta}{R^2}$

$F_x = 2\left(\dfrac{Gm\lambda}{R}\right)\displaystyle\int_0^{\pi/2}\cos\theta\,d\theta = \dfrac{2}{\pi}\dfrac{GmM}{R^2}$

(b) $X_{CM} = \displaystyle\int \dfrac{x\,dM}{M}$, $\qquad Y_{CM} = 0$ BY SYMMETRY

$X_{CM} = 2\displaystyle\int_0^{\pi/2}\dfrac{(R\cos\theta)\lambda R\,d\theta}{M} = \dfrac{2R}{\pi}$

(c) $F = \dfrac{GMm}{(X_{CG})^2} = \dfrac{2}{\pi}\dfrac{GmM}{R^2}$

$X_{CG} = R\sqrt{\dfrac{\pi}{2}}$, $\qquad Y_{CG} = 0$.

GRAVITATION
SATELLITE MOTION

4-64

The figure depicts Saturn and Uranus travelling clockwise around the sun in nearly circular orbits whose radii happen to be almost exactly in the ratio 2 to 1, i.e. $R_{uranus} = 2.01\, R_{saturn}$.
At a certain time, say t = 0, they are positioned as shown, making a straight line with the sun (this is called inferior conjunction). Fifteen years later, Saturn has moved to the opposite side of its orbit, as shown. <u>Where is Uranus then??</u>
Answer by marking a clear dot on Uranus' orbital circle, and show your calculations and reasoning below.

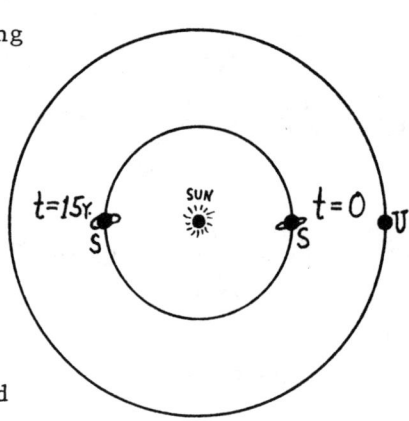

****

Use Kepler's 3rd Law: $T^2 \propto R^3$: then

$$\left(\frac{T_{URANUS}}{T_{SATURN}}\right)^2 = \left(\frac{R_{URANUS}}{R_{SATURN}}\right)^3 = \left(\frac{2}{1}\right)^3 = 8$$

$$T_{URANUS} = \sqrt{8}\, T_{SATURN} = 2.83\, T_{SATURN}$$

So if Saturn's gone 180°, Uranus' gone $\frac{180°}{2.83} = \underline{\underline{64°}}$

So the picture <u>at t = 15 YR</u>:

4-65

(a) State Newton's law of universal gravitation, which refers to the mutual force between masses.

(b) Galileo asserted that all objects, regardless of their masses, will fall to earth in the same time when dropped from the same height (neglecting air resistance). That is, show using Newton's law of gravitation and Newton's second law of motion that an object near the surface of the earth falls with acceleration g, whatever its mass.

(c) Show that, given the distance r_{es} between the earth and the sun and the speed of the earth in its orbit about the sun, you can determine the mass of the sun. Take the earth orbit to be a circle, with the sun fixed at the center of the circle. (HINT: Use Newton's law of gravitation.)

***

(a) $F = G\dfrac{mm'}{r^2}$ $\left[\vec{F} = -G\dfrac{mm'}{r^2}\hat{r}\right]$

(b) $F = ma$
$\overset{a}{=} W = mg$, when Newton's second law is applied to the particular case of a mass m under the influence of the Earth's gravity, say.
$\parallel$
$G\dfrac{mM_E}{R_E^2} = mg$, taking the object to be near the surface of the earth.

$\therefore g = G\dfrac{M_E}{R_E^2}$. That is, $g \neq g(m)$!

(c) Assuming uniform circular motion with the sun at a fixed center and earth at the distance R_{ES},

$F = ma$

$G\dfrac{m_E M_S}{R_{ES}^2} = m_E \dfrac{v^2}{R_{ES}}$

$\therefore M_S = \dfrac{v^2 R_{ES}}{G}$,

where V is the orbital speed of the earth.

$V_E = \left(\dfrac{\text{circumference of earth orbit}}{\text{period (1 year)}}\right)$

$= \dfrac{2\pi R_{ES}}{T(1yr)} = 3.0 \times 10^4 \text{ m/s}$

$R_{ES} = 93 \times 10^6 \text{ mi} = 1.5 \times 10^{11} \text{ m}$

$\therefore M_S = 2 \times 10^{30} \text{ kg}$

4-66

A space adventurer discovers a small planet while exploring a new star system. He puts his ship in circular orbit around the planet and finds that he travels once around the planet in 11 hours. With his navigational equipment he determines that the radius of his orbit is 8000 km. What is the mass of the planet?

****

Describing problem:

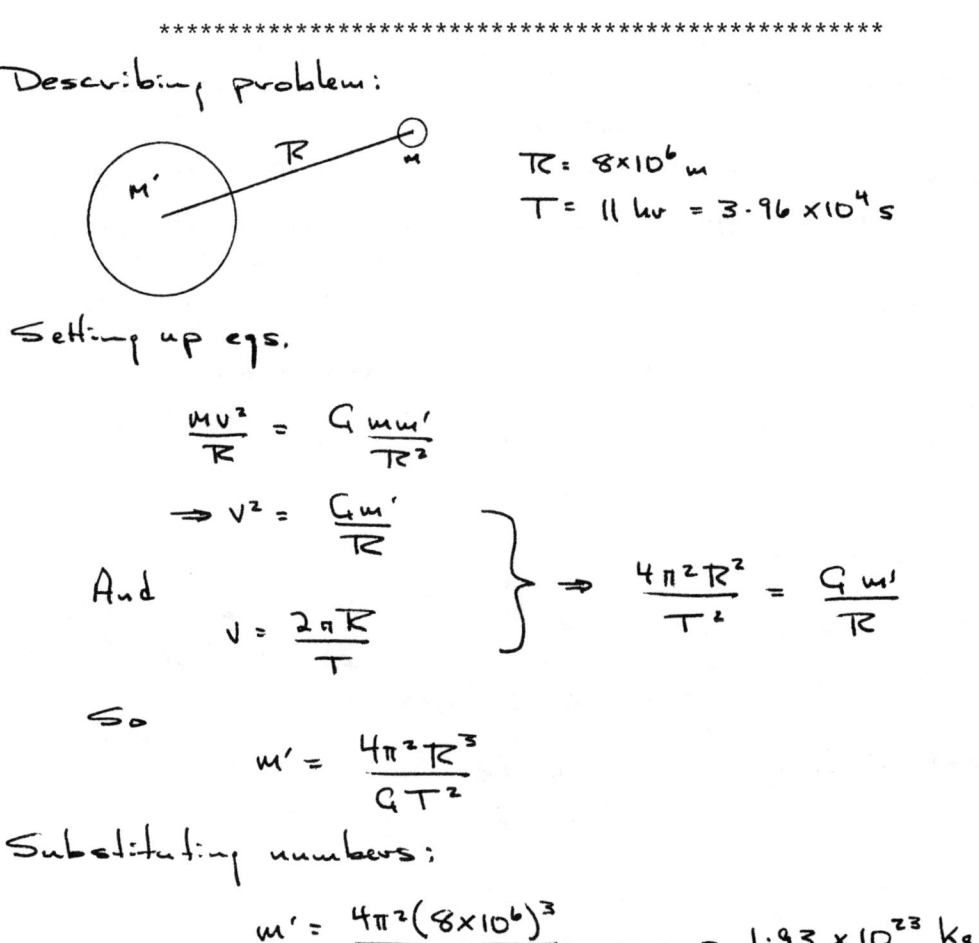

$R = 8 \times 10^6$ m
$T = 11$ hr $= 3.96 \times 10^4$ s

Setting up eqs.

$$\frac{mv^2}{R} = \frac{Gmm'}{R^2}$$

$$\Rightarrow v^2 = \frac{Gm'}{R}$$

And

$$v = \frac{2\pi R}{T}$$

$$\Rightarrow \frac{4\pi^2 R^2}{T^2} = \frac{Gm'}{R}$$

So

$$m' = \frac{4\pi^2 R^3}{GT^2}$$

Substituting numbers:

$$m' = \frac{4\pi^2 (8 \times 10^6)^3}{(6.67 \times 10^{-11})(3.96 \times 10^4)^2} = \underline{\underline{1.93 \times 10^{23}}} \text{ kg}$$

Satellite Motion / 135

4-67

A 100-kg satellite executes a circular orbit 1000 km above the earth's surface. The radius of the earth is 6370 km. Find the satellite's acceleration and the time needed to complete one orbit.

***

This starts out looking like a problem where not enough information is given. But, as is often the case, a bit of thought and some manipulation yield the necessary data. First off, you know the force on the satellite is given by Newton's law of universal gravitation:

$$F = \frac{G M_e m_s}{R_o^2},$$

where M_e = mass of earth
m_s = satellite mass
R_o = orbital radius
G = grav. constant

But neither G nor M_e are given! But — you know that the acceleration of gravity on the earth's surface = g = 9.8 m/s². This acceleration is the gravitational force (weight) $\frac{G M_e m}{R_e^2}$ divided by the mass m (R_e = earth radius). This means that $g = G M_e / R_e^2$ or $g R_e^2 = G M_e$. So the force on the satellite is just $\left(\frac{g R_e^2}{R_o^2}\right) m_s$, and the satellite's acceleration is $\left(\frac{R_e^2}{R_o^2}\right) g = \frac{(6370 \text{ km})^2}{(6370 + 1000 \text{ km})^2} g$

(direction of a: toward earth)

$$\boxed{a = 7.32 \text{ m/s}^2}$$

But this acceleration is the satellite's centripetal acceleration = $v^2/R_o \Rightarrow v^2 = g R_e^2 / R_o$; and v = distance/time = $\frac{\text{circumference}}{\text{period}} = 2\pi R_o / \tau$

$$\Rightarrow \frac{4\pi^2 R_o^2}{\tau^2} = \frac{g R_e^2}{R_o} \Rightarrow \tau^2 = 4\pi^2 R_o^3 / g R_e^2$$

$$\Rightarrow \boxed{\tau = 6300 \text{ sec } (1.75 \text{ hours})}$$

(Note that the mass of the satellite is irrelevant for this problem.)

4-68

The period of revolution of the moon around the earth (ignoring the earth's rotation about its own axis) is approximately 27.3 days. The average distance between the center of the moon and the center of the earth is approximately 3.8×10^8 meters.

Calculate the mass of the earth.

****

Sum the forces acting on the moon along a radial direction passing between the earth-moon centers:

$$\Sigma F = \frac{G M_e M_m}{r^2} = M_m a$$

$$\therefore \frac{G M_e M_m}{r^2} = M_m \frac{v^2}{r}$$

$$M_e = \frac{v^2 r}{G}$$

$$= \frac{\left(\frac{2\pi r}{T}\right)^2 r}{G}$$

$$= \frac{4\pi^2 r^3}{G T^2}$$

$$= \frac{4\pi^2 (3.8 \times 10^8)^3}{6.67 \times 10^{-11} (27.3 \times 24 \times 3600)^2}$$

$$\approx 6 \times 10^{24} \text{ kg.}$$

Two small masses revolve about each other in a region of space far from other objects. Each mass moves along a circular path. (a) Find the radius of each circular orbit. (b) Find the period of revolution.

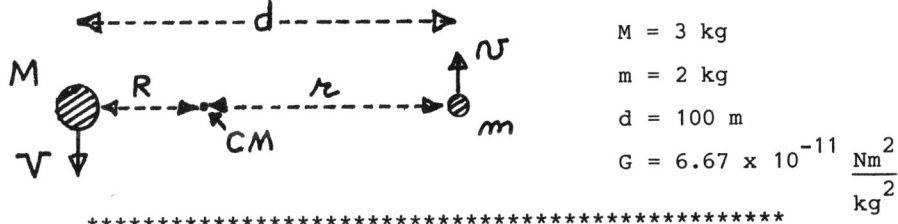

M = 3 kg
m = 2 kg
d = 100 m
G = 6.67 x 10^{-11} $\frac{Nm^2}{kg^2}$

****

(a) PLACE THE ORIGIN AT THE CENTER OF MASS.

$$X_{CM} = \frac{\sum m_i r_i}{\sum m_i} = \frac{mr + M(-R)}{m + M} = 0$$

$$mr = MR, \quad d = r + R, \quad d = r + \frac{m}{M}r = r\left(1 + \frac{m}{M}\right)$$

$$r = \frac{d}{\left(1 + \frac{m}{M}\right)} = \frac{100}{\left(1 + \frac{2}{3}\right)}, \quad r = 60 m, \quad R = 40 m$$

(b) THE GRAVITATIONAL FORCE ACTING ON EITHER OBJECT IS THE CENTRIPETAL FORCE REQUIRED FOR CIRCULAR MOTION.

$$\frac{GMm}{d^2} = \frac{mv^2}{r}, \quad \frac{GM}{d^2} = \frac{1}{r}\left(\frac{2\pi r}{T}\right)^2 = \frac{4\pi^2}{T^2}\left(\frac{d}{1 + \frac{m}{M}}\right)$$

$$T = 2\pi\sqrt{\frac{d^3}{G(M+m)}} = 2\pi\sqrt{\frac{(100)^3}{6.67 \times 10^{-11}(3+2)}}$$

$$T = 3.44 \times 10^8 \, s = 10.9 \text{ YEARS.}$$

4-70

Pluto orbits the sun in an essentially circular orbit of radius 5.9×10^{12} meters with a period of 7.8×10^9 seconds. Find (if possible):

(a) the linear speed;
(b) the angular speed;
(c) the centripetal acceleration;
(d) the rate at which a line from the sun to Pluto sweeps out area;
(e) the mass of the sun;
(f) the angular momentum.
(g) Can you determine the mass of Pluto? If the answer is no, which of the above questions, a to f, can you, likewise, not answer?

****

a) $v = \dfrac{2\pi R}{T} = \dfrac{(6.28)(5.9 \times 10^{12} m)}{7.8 \times 10^9 s} = 4.75 \times 10^3 \, m/s$

b) $\omega = \dfrac{v}{r} = (4.75 \times 10^3) \, m/s \,/\, (5.9 \times 10^{12} m) = .8 \times 10^{-9} \, \dfrac{radians}{sec.}$

but also $\omega = \dfrac{2\pi}{T} = 6.28 / (7.8 \times 10^9 s) = .8 \times 10^{-9} \, s^{-1}$

c) $a_c = \dfrac{v^2}{r} = (4.75 \times 10^3 \, m/s)^2 / (5.9 \times 10^{12} m) = 3.8 \times 10^{-6} \, m/s^2$

but also $a_c = \omega^2 r = (.8 \times 10^{-9} \, s^{-1})^2 (5.9 \times 10^{12} m)$

which gives, again, $a_c = 3.8 \times 10^{-6} \, m/s^2$

d) $\dfrac{dA}{dt} = \dfrac{1}{2} R \dfrac{(R d\theta)}{dt} = \dfrac{1}{2} R^2 \omega$ since, when the time interval is very small, the wedge shaped piece of pie is a triangle having shorter side $R d\theta$ and longer side R. Thus

$\dfrac{dA}{dt} = \dfrac{1}{2}(5.9 \times 10^{12} m)^2 (.8 \times 10^{-9} s^{-1}) = 13.9 \times 10^{15} \, m^2/s$

e) The centripetal force acting on Pluto is also the gravitational force of attraction to the sun so $G \, M_s M_p / R^2 = m_p v^2 / R$

OR $\quad M_s = \dfrac{Rv^2}{G} = \dfrac{(5.9 \times 10^{12}\,m)(4.75 \times 10^3\,m/s)^2}{6.67 \times 10^{-11}\,\frac{New \cdot m^2}{Kg^2}}$

$M_s = 20 \times 10^{29}\,Kg.$

f) g) $\ell = m_p v r = 2 m_p \dfrac{dA}{dt}$ and cannot be determined since m_p cannot be determined from the information given.

4-71

(a) The moon's orbital period is 27.32 days, and its orbit radius is 3.844E8 m. Compute the magnitude of the earth's gravitational field out where the moon is located.
(b) Show that your answer to (a) is consistent with inverse-square-law distance dependence. The radius of the earth is 6.378E6 m.

***

(a) $T = 2\pi \sqrt{\dfrac{r}{g}}$ relates period and radius for circular orbits.

Thus $g = \dfrac{4\pi^2 r}{T^2} = \dfrac{4\pi^2 \times 3.844 \times 10^8\,m}{(27.32\,days \times 86400\,sec/day)^2} = \underline{2.72 \times 10^{-3}\,m/sec^2}$

(b) $g_M = g_E \left(\dfrac{r_E}{r_M}\right)^2$ should be true if the earth's gravity decreases in proportion to distance-squared (distance from the earth's center).

$g_E \left(\dfrac{r_E}{r_M}\right)^2 = 9.8 \times \left(\dfrac{6.378 \times 10^6}{3.844 \times 10^8}\right)^2 = 2.70 \times 10^{-3}\,m/sec^2$, which agrees well with answer (a).

MOTION IN A VERTICAL CIRCLE

4-72

A roller coaster vehicle has a mass of 700 kg. If the vehicle has a speed of 25 m/s at point A, what is the force of the track on the vehicle at this point?

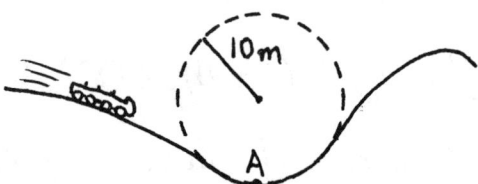

***

Free-body diagram:

("normal" force)
$N = ?$
$W = mg$

$F = ma \Rightarrow N - W = ma$

$$N - mg = \frac{mv^2}{r}$$

$$N = m\left(g + \frac{v^2}{r}\right)$$

$$N = (700 \text{ kg})\left[\frac{9.8 \text{ m}}{\text{sec}^2} + \frac{(25 \text{ m}/\text{sec})^2}{(10 \text{ m})}\right]$$

$$N = 5.1 \times 10^4 \text{ Newtons}$$

4-73

A 49 n child on a swing is swinging in an arc of 3m radius. If the horizontal speed at the lowest point is 3 m/s, find the tension in the rope.

****

$T - mg = ma$

$T - mg = mv^2/R$

$T = mv^2/R + mg$

$T = \dfrac{49}{9.8} \times \dfrac{9}{3} + 49$

$T = 57 \, n$

RELATIVE VELOCITY

4-74

You are flying a level course in an airplane. Your instruments show an airspeed of 200 mph and a direction due North. If there is a wind blowing toward the North-West at 40 mph, what is the velocity and direction of the airplane's motion with respect to the earth?

****

$$\vec{v}_{\text{PLANE TO EARTH}} = \vec{v}_{\text{PLANE TO AIR}} + \vec{v}_{\text{AIR TO EARTH}}$$

$$(v_W, v_N) = (0, 200) + (40 \cos 45°, 40 \sin 45°)$$

$$(v_W, v_N) = (20\sqrt{2}, 200 + 20\sqrt{2}) \text{ MPH}$$

$$= (28.3, 228.3) \text{ MPH}$$

THEN $v = \sqrt{(28.3)^2 + (228.3)^2}$

$$v = 230 \text{ MPH}$$

AND $\theta = \arctan \dfrac{28.3}{228.3}$

$$\theta = 7.1° \text{ WEST OF NORTH}$$

Relative Velocity / 143

4-75

A large and unpleasant bird is headed west at an altitude of 100 meters and an airspeed of 2.0 m/s. The wind is from 45° east of north at a speed of 1.0 m/s, so the bird is not moving exactly west with respect to the ground. If the bird releases a 300 gram mass directly over the origin (x=0, y=0) where will the mass hit the ground? Neglect air resistance, and assume that the bird simply drops the mass.

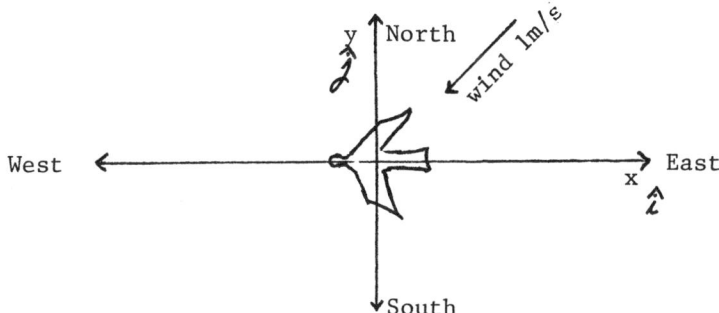

****

$$\vec{v}_{\text{bird wrt ground}} = \vec{v}_{\text{bird wrt air}} + \vec{v}_{\text{air wrt ground}}$$

(wrt = with respect to)

$$= (-2 \text{ m/s } \hat{i}) + (-1 \text{ m/s} \cos 45° \hat{i} - 1 \text{ m/s} \sin 45° \hat{j})$$

$$= -2.71 \text{ m/s } \hat{i} - 0.71 \text{ m/s } \hat{j}$$

This is also the velocity of the mass when the bird releases it. Gravity accelerates the mass in the vertical (z or $\hat{k}$) dimension, but the horizontal components remain unchanged. We must find the time of the fall to see how far the mass moves.

mass starts 100m up & ends on ground.

$$z = z_0 + v_{0z} t + \tfrac{1}{2} a t^2$$
$$0 = 100 \text{m} + 0 - \tfrac{1}{2}(9.8 \text{ m/s}^2) t^2$$
$$t = \sqrt{\frac{2(100 \text{m})}{9.8 \text{ m/s}^2}} = 4.5 \text{ s}$$

Now calculate its horizontal position at this time.

$$\vec{r} = \vec{r_0} + \vec{v_0} t + \tfrac{1}{2}\vec{a}t^2 \quad \text{starts over origin; no horizontal } \vec{a}.$$

$$= -12.2 \text{ m } \hat{i} - 3.2 \text{ m } \hat{j}$$

OR   12.6 m from origin at 15° South of West

4-76

An empty truck, 6 m long from front to rear axle is at rest with the front tires on a scale, A, and its rear tires on scale B. Scale A reads 40,000 Newtons and scale B reads 20,000 Newtons.

(a) Where is the center of mass of the empty truck?
(b) A 10,000 Newton load is now placed 1 m from the rear axle. Find the readings of scale A and B.
(c) When the truck is moving on the road it is noticed that, when the brakes are applied, it takes 3 seconds for the load to slide 5 meters across the <u>frictionless</u> truck bed. If the center of mass of the loaded truck is 1 m above the front and rear axles, find the weight supported by the front and the rear axles when the brakes are applied if the load is tied down.

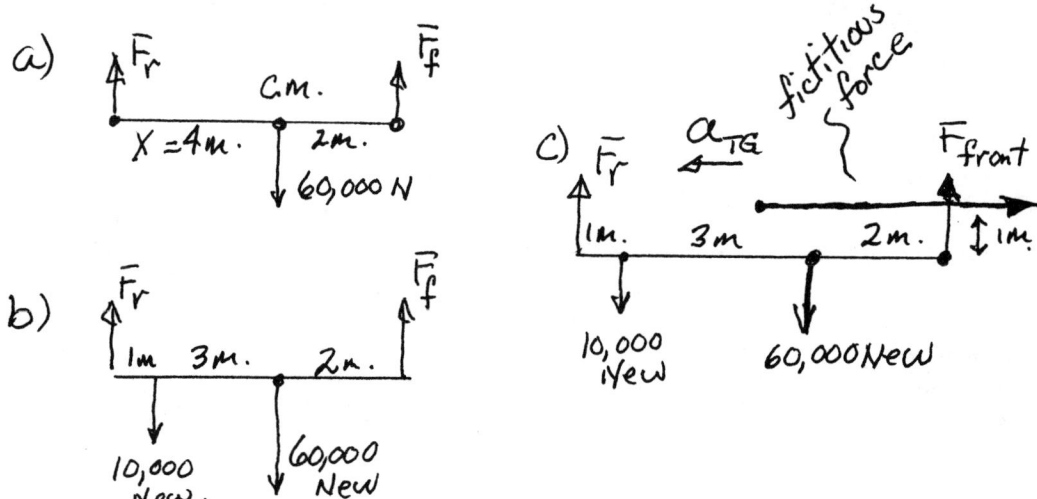

***

a) Calculating torques about the rear axle gives

$(x)(6 \times 10^4 \text{ New}) = (6m) F_f = 240,000; \quad x = 4m.$

b) Calculating torques about the front axle gives

$6 F_r = 2(6 \times 10^4 N) + 5(1 \times 10^4); \quad F_r = 28,333 N$

Hence $F_f = 41,667$ since $F_r + F_f = 70,000 N$

c) Let's work in a frame of reference attached to the Truck. We need to calculate the acceleration of the ground with respect to the Truck, or vice versa. If we use a double subscript notation in which a_{LT} is the acceleration of the load across the Truck (or as seen from the Truck), etc., then there is a rule which says that $a_{LT} = a_{LG} + a_{GT}$ where $G \Rightarrow$ ground.

Now since $a_{LG} = 0$, $a_{LT} = a_{GT} = -a_{TG}$

But $5m = \frac{1}{2}(a_{LT})(3s)^2$ so $a_{LT} = 1.11 \, m/s^2$

The fictitious (or inertial) force acting in the Truck frame acts at the center of mass and points opposite to the acceleration vector a_{TG} and its magnitude is given by $|\vec{F}| = m_T |\vec{a}_{TG}|$

that is, $(7000 \, kg)(1.11 \, m/s^2) = 7,800$ New.

Taking Torques about the rear axle gives
$$(1)(10,000) + (4)(60,000) + (1)(7800) = 6 \, \overline{F}_f$$
$$\overline{F}_f = 43,000 \text{ New.}$$

Hence $\overline{F}_r = 27,000$ since $\overline{F}_f + \overline{F}_r = 70,000$ New.

4-77

Given: Velocity $\vec{V}_1 = 2\vec{i} + 3\vec{j} - 6\vec{k}$ (m/s) and velocity $\vec{V}_2 = 3\vec{i} - 2\vec{j} + 2\vec{k}$ (m/s)

(a) Find the velocity $\vec{V} = \vec{V}_1 + \vec{V}_2$ (m/s) and the velocity $\vec{V}' = \vec{V}_1 - \vec{V}_2$.
(b) Find the angle enclosed by velocity vectors $\vec{V}_1$ and $\vec{V}_2$
    [Hint: Use the scalar product]
(c) Imagine $\vec{V}_1$ to be the velocity of a boat as measured relative to the stream through which it is passing, and $\vec{V}_2$ the velocity of the stream.
  (i) Find the velocity of the boat, as measured by an observer standing on the shore.
  (ii) Find the actual displacement $\Delta \vec{r}$ of the boat in 50 seconds, as determined by the observer standing on the shore.
  (iii) Find the distance the boat travels, as determined by the observer standing on the shore.

****

(a) $\vec{V} = \vec{V}_1 + \vec{V}_2 = (2+3)\vec{i} + (3-2)\vec{j} + (-6+2)\vec{k}$

$$\vec{V} = 5\vec{i} + \vec{j} - 4\vec{k} \quad m/s$$

$\vec{V}' = \vec{V}_1 - \vec{V}_2 = (2-3)\vec{i} + (3-(-2))\vec{j} + (-6-(+2))\vec{k}$

$$\vec{V}' = -\vec{i} + 5\vec{j} - 8\vec{k} \quad m/s$$

(b) $\vec{V}_1 \cdot \vec{V}_2 = V_1 V_2 \cos\theta$; θ, the angle enclosed by $\vec{V}_1$ and $\vec{V}_2$

Also $\vec{V}_1 \cdot \vec{V}_2 = V_{1x} V_{2x} + V_{1y} V_{2y} + V_{1z} V_{2z}$

Equating,

$V_1 V_2 \cos\theta = V_{1x} V_{2x} + V_{1y} V_{2y} + V_{1z} V_{2z}$

$$\therefore \theta = \cos^{-1} \frac{V_{1x} V_{2x} + V_{1y} V_{2y} + V_{1z} V_{2z}}{V_1 V_2}$$

where $V_1 = \sqrt{V_{1x}^2 + V_{1y}^2 + V_{1z}^2}$; $V_2 = \sqrt{V_{2x}^2 + V_{2y}^2 + V_{2z}^2}$

$$\theta = \cos^{-1} \frac{(2)(3) + (3)(-2) + (-6)(2)}{\sqrt{4+9+36}\sqrt{9+4+4}} = \cos^{-1}(-0.416)$$

$$\theta = 114.6°$$

(c) As seen in the accompanying diagram, at any given instant the boat has position given by $\vec{r}'$ or $\vec{r}$, depending on whether the observer is O' or O, i.e., moving with the stream or standing on the shore, respectively.

(Of course O and O' disagree, for at the given instant they are at different positions, as described by $\vec{r_2}$)

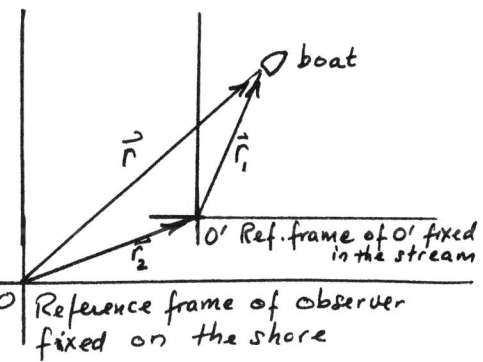

(i) These positions will change from instant to instant, since both stream and boat are moving.

Since $\vec{r} = \vec{r_1} + \vec{r_2}$, $\dfrac{d\vec{r}}{dt} = \dfrac{d\vec{r_1}}{dt} + \dfrac{d\vec{r_2}}{dt}$ or

$\vec{U} = \vec{U_1} + \vec{U_2}$

Then the velocity of the boat as seen by the shore-based observer O is

$\vec{U} = \vec{U_1} + \vec{U_2} = (5\hat{\imath} - \hat{\jmath} - 4\hat{k})$ m/s (See Part(a))

(ii) $\Delta \vec{r} = \vec{V} \Delta t$ since $\vec{U}$ = constant.

$\Delta \vec{r} (\Delta t = 50 s) = (5\hat{\imath} - \hat{\jmath} - 4\hat{k}) \left(\dfrac{m}{s}\right) \times 50 \, (s)$

$\Delta \vec{r} = (250 \hat{\imath} - 50 \hat{\jmath} - 200 \hat{k})$ m

(iii) ∴ Distance travelled $\Delta r = \sqrt{(\Delta x)^2 + (\Delta y)^2 + (\Delta z)^2}$

$= \sqrt{(250)^2 + (50)^2 + (200)^2}$

$\Delta r = 324$ (m)

4-78

A pilot must set a course so that his airplane flies due east. The plane can develop a speed of 250 km/hr relative to the air and there is a wind blowing from the north at 50 km/hr.
a) Take the x-axis pointing east and the y-axis pointing north. Write down the components of the wind's velocity, $\vec{w}$, and the plane's velocity, $\vec{v}_g$, relative to the ground.
b) Draw a clear diagram relating $\vec{w}$, $\vec{v}_g$, and $\vec{v}_{air}$, the plane's velocity relative to the air.
c) How long does it take for the plane to reach its target, 100 km to the east?

****

a) $\vec{w} = (0, -50)$ km/hr $\qquad \vec{v}_g = (v_g, 0)$ km/hr

b)

$V_{air} = 250$ km/hr, $w = 50$ km/hr, v_g

c) $\therefore v_g = \sqrt{V_{air}^2 - w^2} = \sqrt{(250)^2 - (50)^2} = 245$ km/hr

$\therefore$ time $= \dfrac{\text{distance}}{\text{speed}} = \dfrac{100 \text{ km}}{245 \text{ km/hr}} = .41$ hrs

$\therefore$ time for journey $= 24.5$ mins.

5
WORK AND ENERGY

WORK DONE BY A CONSTANT FORCE

■■■ 5-1

Two masses, m_1 and m_2, are connected by a light string, and are being pulled a distance d along a surface at a constant speed by a force F being applied at an angle α above the horizontal to mass m_1. The coefficient of kinetic friction between blocks and surface is μ.
(a) For each mass, set up a diagram showing all forces acting, and write the corresponding vector equations.   (b) In terms of m_1, m_2, α, g, and μ, determine the magnitudes of force F and the tension T in the connecting string.   (c) Let m_1= 5kg, m_2= 3kg, α=30°, μ=0.2, v= 6m/sec, and d= 8m. Find the numerical values of F and T, the work done by F, and the instantaneous power expended by F in this process.

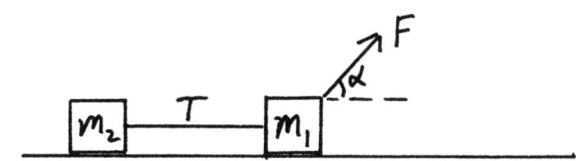

****

(a)
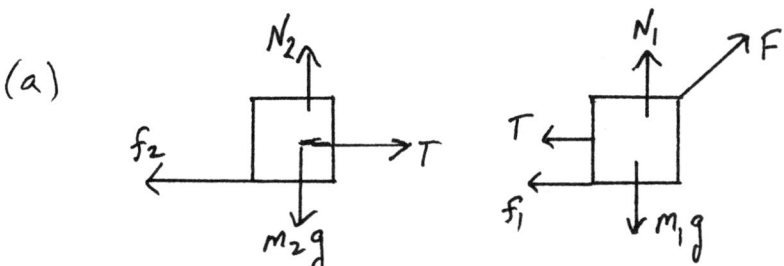

149

For mass m_1: $F\cos\alpha - f_1 - T = 0$
$F\sin\alpha + N_1 - m_1 g = 0 \qquad f_1 = \mu N_1$

For mass m_2: $T - f_2 = 0 \qquad f_2 = \mu N_2 \qquad N_2 - m_2 g = 0$

(b) T is determined from the equations for m_2:
$T = f_2 = \mu N_2 = \mu m_2 g$

F is then determined from the equations for m_1:
$N_1 = m_1 g - F\sin\alpha \qquad f_1 = \mu(m_1 g - F\sin\alpha)$
$F\cos\alpha = f_1 + T = \mu(m_1 g - F\sin\alpha) + \mu m_2 g$
$F\cos\alpha + \mu F\sin\alpha = \mu g (m_1 + m_2)$
$F = \mu g (m_1 + m_2) / (\cos\alpha + \mu \sin\alpha)$

$F = (.2)(9.8)(5+3) / (\cos 30° + .2 \sin 30°) = 16.2 N$

$T = (.2)(3)(9.8) = 5.88 N$

$W = (F\cos\alpha) d = (16.2)(\cos 30°)(8) = 112 J$

$P = (F\cos\alpha) v = (16.2)(\cos 30°)(6) = 84.2 \text{ Watts}$

5-2

The 15 kg block is pulled 5 m up the incline as shown. The coefficient of friction between the block and the incline is 0.4.
(a) What is the work done by the 500 N force?
(b) What is the work done by the weight of the block?
(c) What is the work done by the friction force?
(d) What is the work done by the 407 N normal force?

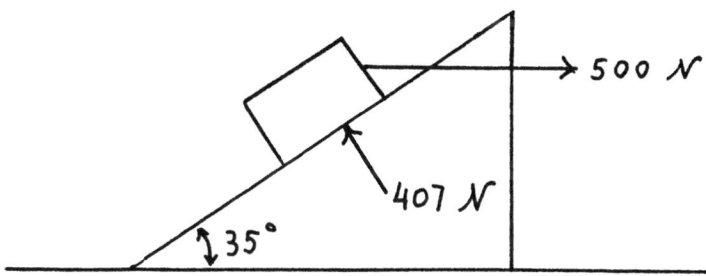

****

(a) Using the equation $W = Fs\cos\theta$ we obtain

$$W = (500)(5)\cos 35° = \underline{\underline{2048 \text{ J}}}$$

(b) The weight = mg = 15(9.8), hence

$$W = 15(9.8)(5)\cos 125° = \underline{\underline{-421.6 \text{ J}}}$$

(c) The friction force $f = \mu N = 0.4(407)$ and

$$W = 0.4(407)(5)\cos 180° = \underline{\underline{-814 \text{ J}}}$$

(d) Since the normal force is perpendicular to the displacement $W = 407(5)\cos 90° = \underline{\underline{0}}$

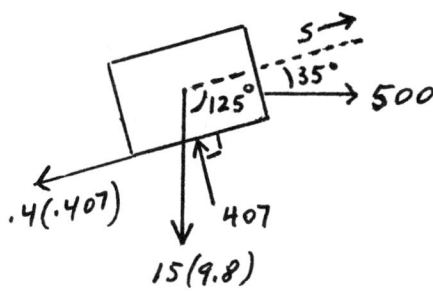

5-3

A winch is slowly unwinding to allow a mass (m = 100 kg) to move down a slope at a constant acceleration of 0.3 m/sec². The slope is 6 m long and makes an angle of 35° with respect to the horizontal. There is a coefficient of friction μ = 0.25 between the block and the slope. (a) How much work is done by the frictional force in lowering the block all the way to the bottom of the slope? (b) How much work is done by gravity? (c) How much work is done by the winch?

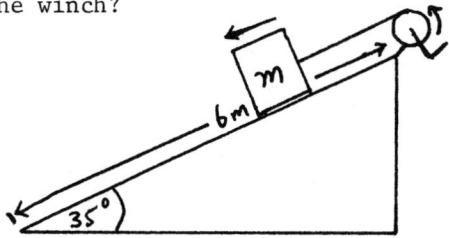

****

(a) Draw a free-body diagram for the block:

T = tension in string
f = force of friction
N = Normal force
mg = weight of block

$$W = \vec{F} \cdot \vec{D}$$

$f = \mu N$; Using Newton's Second Law and taking the force components that are ⊥ to the slope
$\Sigma F_\perp = 0$
$N - mg\cos 35° = 0$
$N = mg\cos 35°$
$f = (.25)(100)(9.8)\cos 35° = 200.7$ N
$W = fD \cos 180° = (200.7)(6)(-1) = \underline{\underline{-1204 \text{ J}}}$

(b) $W = \vec{F} \cdot \vec{D} = mgD \cos 55°$

$= (100)(9.8)(6) \cos 55° = \underline{\underline{3373 \text{ J}}}$

(c) $W = \vec{F} \cdot \vec{D} = TD \cos 180°$; Using Newton's Law and taking the components ∥ to the slope we get T
$\Sigma F_\parallel = ma$
$mg\sin 35° - f - T = ma$
$T = mg\sin 35° - f - ma = 331.4$ N
$W = (331.4)(6)(-1) = \underline{\underline{-1988.4 \text{ J}}}$

Work Done By A Constant Force / 153

5-4

You are using a motor which produces one horsepower (746 watts) to drag a 100 kg block up a frictionless 40° incline at constant speed. What is the maximum speed at which the motor can pull the block up the incline? How much work does the motor do in getting the block 3.00 meters above the ground?

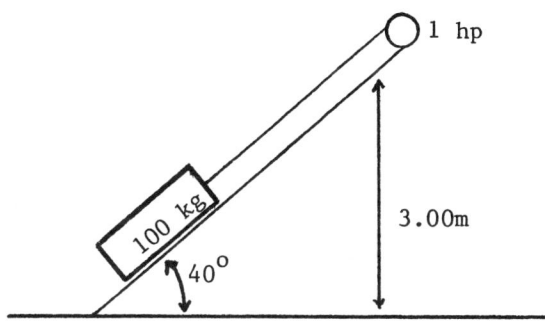

****

First Part: Want to use
$$P = \vec{F} \cdot \vec{v}$$
Must get $\vec{F}$:

x-dimension: $\Sigma F = ma$

$F - mg \sin\theta = 0$ [constant velocity means $a = 0$]

$F = mg \sin\theta$

So $P = (mg \sin\theta) v$

$v = \dfrac{P}{mg \sin 40°} = 1.2 \text{ m/s}$

Last Part: No friction; all forces are conservative, so work just depends on the net result — not the path taken.

$W = mgh = (100 \text{ kg})(9.8 \text{ m/s}^2)(3.00 \text{ m})$

$W = 2900 \text{ J}$

5-5

A cord is used to lower a block of weight 4.00 newtons by a distance of 2.00 meters at a constant downward acceleration of 1/4 of the acceleration of gravity. Find the force exerted on the block by the cord and the work done by the cord on the block during this process. Assume that the only forces acting on the block are its weight and the upward pull of the cord.

****

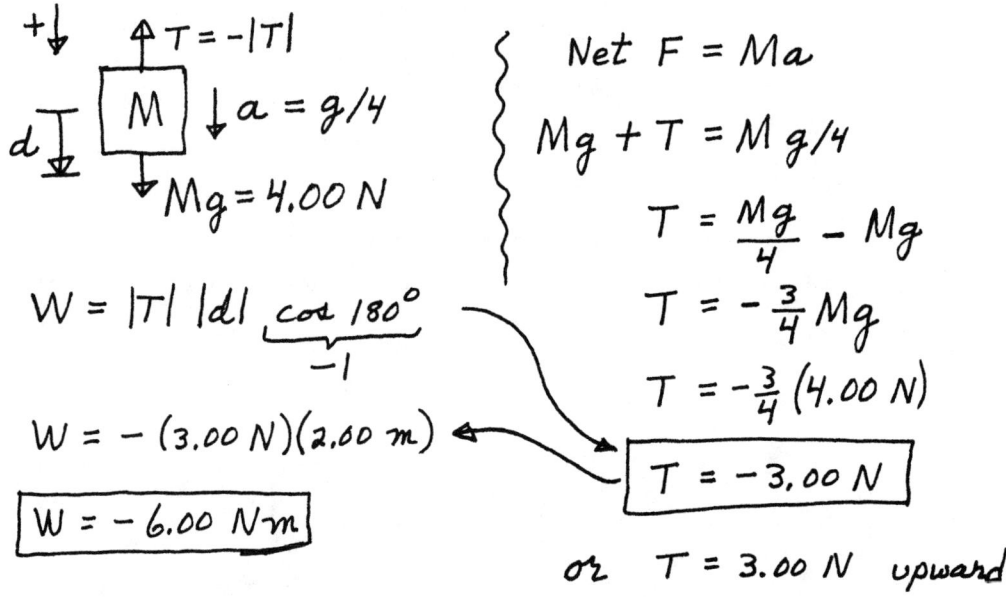

$$W = |T| |d| \underbrace{\cos 180°}_{-1}$$

$$W = -(3.00 \text{ N})(2.00 \text{ m})$$

$$\boxed{W = -6.00 \text{ N·m}}$$

$$\text{Net } F = Ma$$
$$Mg + T = Mg/4$$
$$T = \frac{Mg}{4} - Mg$$
$$T = -\frac{3}{4} Mg$$
$$T = -\frac{3}{4}(4.00 \text{ N})$$
$$\boxed{T = -3.00 \text{ N}}$$

or $T = 3.00$ N upward

5-6

A box whose weight is 80 nt is pulled up along the inclined plane a distance of 20 m by the applied force of 100 nt as shown. The coefficient of kinetic friction is 0.22. Calculate the change in the kinetic energy of the box.

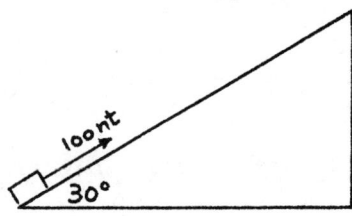

****

The input amount of energy into the system:
   (100)(20) nt·m .

This energy is divided among friction loss, potential energy increase and kinetic energy change, i.e.,

The change in kinetic energy $= (100)(20) - (0.22)(80 \cos 30°) \times (20) - (80)(20 \sin 30°) \cong 895.2$ nt·m (or joules) ←

WORK DONE BY A VARIABLE FORCE

━━ 5-7

The force required to pull back a compound archery bow varies approximately as shown in the graph below. How much energy is stored in the bow when drawn to 30 inches?

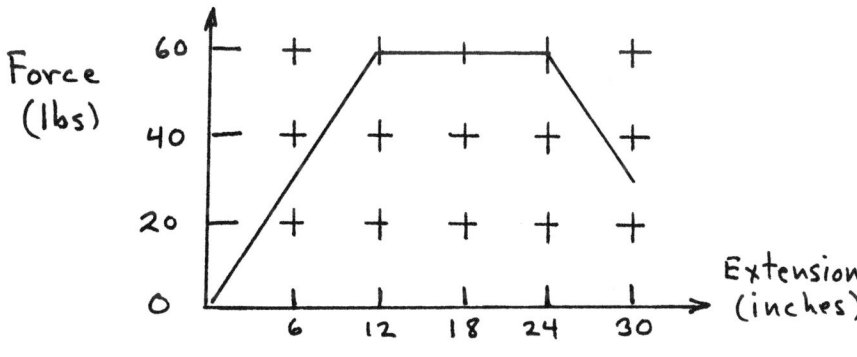

****

The energy stored is equal to the work done pulling the string back, which is the area between the force curve and the force = 0 axis. The area is that of a triangle (0" to 12") plus a rectangle (12" to 24") plus a trapezoid (24" to 30").

Work $= \frac{1}{2}(60 lb)(1 ft) + (60 lb)(1 ft) + \frac{1}{2}(60 lb + 30 lb)(0.5 ft)$

$= \underline{\underline{112.5 \text{ ft lbs}}}$

5-8

Calculate the work done by the force shown below as it moves a body a distance of four feet.

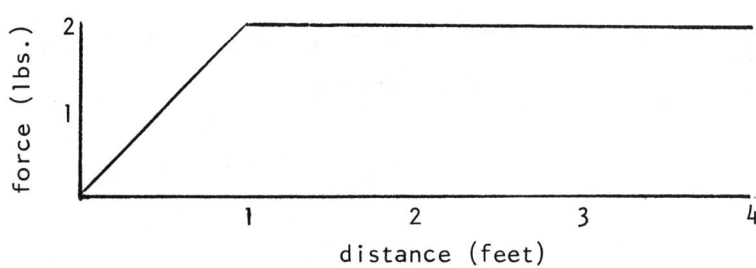

****

$$W = \int \vec{F} \cdot d\vec{r}$$

$$= \int F \, dx$$

$$= \int_0^1 F \, dx + \int_1^4 F \, dx$$

From the graph, one notes that from $x = 0$ to 1 foot
$$F = 2x$$

while from $x = 1$ to 4 ft.
$$F = 2$$

$$\therefore W = \int_0^1 2x \, dx + \int_1^4 2 \, dx$$

$$= 2 \left.\frac{x^2}{2}\right|_0^1 + 2x \Big|_1^4$$

$$= 1 + (8 - 2)$$

$$= 7 \text{ ft. lbs.}$$

Calculate the work that the force, F(x), does as it moves a particle from x = 1 meter to x = 3 meters.

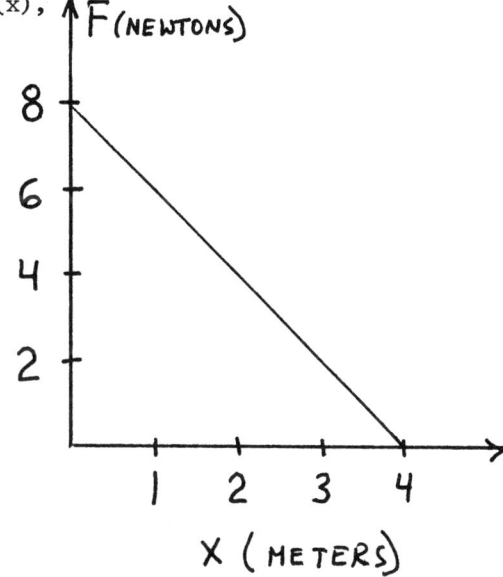

*****

The quickest way is to use the geometric significance of the integral as area under the curve, i.e.

$$W = \int_1^3 F\,dx = \triangle + \square$$

$$= \tfrac{1}{2}(4N.)(2m.) + (2N.)(2m)$$

$$= 4\text{ J.} + 4\text{ J.}$$

$$\boxed{W = 8 \text{ JOULES}}$$

But you can also do it by noting that $F(x) = 8 - 2x$, so that $W = \int_1^3 (8 - 2x)\,dx = [8x - x^2]_1^3 = \underline{8 \text{ JOULES}}\checkmark$

5-10

Physicists believe that protons are made up of three more elementary particles called quarks. Inside a proton a single quark is attracted towards the center of the proton by a force: $F=-(a/r^2+b)$, where a and b are positive constants. Calculate the work done when a quark is moved from a distance r_0, close to the center, to a distance R, far from the center of the proton. What happens as R approaches infinity?

****

$$\text{Work} = \int_{r_0}^{R} F\,dr = -\int_{r_0}^{R}\left[\frac{a}{r^2}+b\right]dr = \left[\frac{a}{r}-br\right]_{r_0}^{R}$$

$$= a\left[\frac{1}{R}-\frac{1}{r_0}\right]-b\left[R-r_0\right]$$

This is the work done by the force. The work that must be done by an external force to cause this movement is the negative of the above expression.

$$\therefore \text{Work needed to move quark} = -a\left[\frac{1}{R}-\frac{1}{r_0}\right]+b\left[R-r_0\right]$$

As $R \to \infty$ this becomes:

$$\text{Work needed to free quark from proton} = \frac{a}{r_0} + \lim_{R\to\infty} bR - br_0 \to \infty$$

Thus, an infinite amount of work is necessary to free the quark — the quark is "confined" to live in the proton — it cannot be freed by any finite amount of energy.

5-11

Consider a force that varies with distance according to the formula F=0.5 X. (The force in newtons is one half the distance to the origin measured along the x-axis in meters.) If an object, acted on by this force, moves along the x-axis from 1.5 m to 2.5 m, how much work does this force do on the object?

****

ON THE F-X GRAPH THE
WORK = AREA
= TRIANGLE 1 - TRIANGLE 2
= $\frac{1}{2}$(2.5m)(0.5 × 2.5N)
  − $\frac{1}{2}$(1.5m)(0.5 × 1.5N)
= 1.0 J

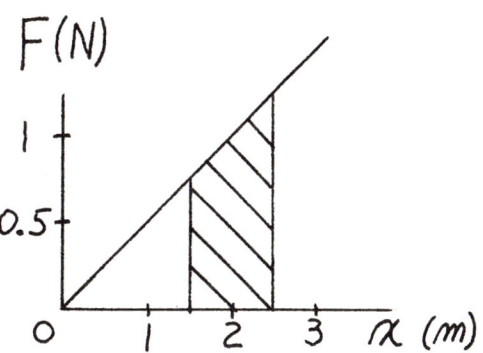

ALTERNATIVELY
$$W = \int F\,dx = \int_{1.5m}^{2.5m} 0.5x\,dx$$

OR W = 1.0 J AS BEFORE

5-12

If the pressure in a balloon is given by p = 30V where V is to be in m^3 and p in N/m^2 find the work done in blowing the balloon up until it has a volume of 0.7 m^3. (Its initial volume is zero.) Hint: plot a pV graph.

****

Describing problem:

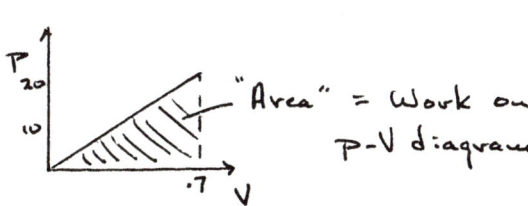

"Area" = Work on p-V diagram

Setting up eqs.

$$W = \tfrac{1}{2} P_f V_f = \tfrac{1}{2}(30 V_f) V_f = \tfrac{1}{2} 30 V_f^2 = 15 V_f^2$$

Substituting numbers:

$$W = 15(.7)^2 = \underline{\underline{7.35\ J}}$$

5-13

The expression for the force on an object as a function of the object's position is:

$$F = 2 + 4x - x^2,$$ where F is in N and x is in m.

a. Find the force on the object 2 m from its equilibrium point.
b. Find the work done in moving the object from $x = 2$ m to $x = 5$ m.

****

a. $F(2) = 2 + 4(2) - (2)^2 = 6\,N.$

b. $W = \int_2^5 F\,dx = \int_2^5 (2 + 4x - x^2)\,dx = 2x + 2x^2 - \frac{x^3}{3} \Big|_2^5$

$\qquad = (10 + 50 - \frac{125}{3}) - (4 + 8 - \frac{8}{3}) = 9\,J.$

KINETIC ENERGY

5-14

A 1 kg ball is launched straight upward with initial speed 50 m/sec, and it reaches a peak altitude of 100 m. For the interval between launch and apogee, compute the following:
(a) Change in kinetic energy.
(b) Work done by the force of gravity.
(c) Work done by the friction-like force of aerodynamic drag.

****

(a) $\Delta KE = KE_2 - KE_1 = 0 - \frac{1}{2} m v_1^2 = -\frac{1}{2} \times 1 \times 50^2 = \underline{-1250\,J}$

(b) $W_{mg} = -mg\,\Delta h = -1 \times 9.8 \times 100 = \underline{-980\,J}$

(force and displacement in opposite directions)

(c) By the work-energy principle, $\Delta KE = W_{net} = W_{mg} + W_f$

(gravity and "friction" are the only forces acting ...)

Thus $W_f = \Delta KE - W_{mg} = -1250 + 980 = \underline{-270\,J}$

5-15

Consider a body of mass 5 kg. placed at rest on a table which has friction. Let a force F = 10N. (as shown) be exerted on the block while it moves a distance of 3 meters. At that point the force is removed and the block is allowed to slide to a stop. How far from its initial point will the block travel. (coefficient of friction = .1)

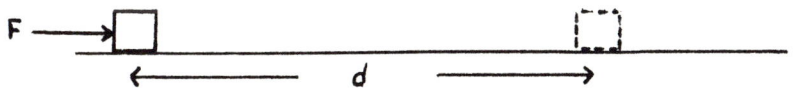

***

First consider the block as it moves from 0 to 3m. Its increase in kinetic energy is equal to the net work.

$$\Delta k = W_{net}$$
$$= \int_0^3 F_{net} \, dx$$

But $F_{net} = 10 - \mu mg$
$$= 10 - .1(5)(9.8)$$
$$= 5.1 \, N$$

$$\therefore \Delta k = \int_0^3 (5.1) \, dx$$
$$= 15.3 \text{ joules}$$

For the region $x = 3m$ to where it stops, all the kinetic energy is expended doing work against the force of friction.

$$W = 15.3 \, j$$
$$\mu mg (d-3) = 15.3 \, j$$
$$d - 3 = \frac{15.3}{\mu mg}$$
$$= \frac{15.3}{.1(5)(9.8)} = 3.1$$

$$\therefore d = 6.1 m.$$

Work And Energy

5-16

A wooden block having mass m = 2.5 kg. moves to the right on a horizontal surface with coefficient of kinetic friction μ_k = 0.20 and, at some instant, speed V_0 = 9.0 m/s

(a) Using the work-energy theorem (which relates work done by the resultant force on an object to the increase of kinetic energy of the object), find how much farther the block moves before stopping.
(b) What is the work done by the normal force in this displacement? What is the work done by the gravity force in this displacement?
(c) Using Newton's second law of motion, find the acceleration of the block.
(d) Find the average power expended in the stopping of the block by the friction force.

****

(a) $W = \Delta K$

$W = \vec{f}_k \cdot \vec{d}$

$\therefore W = -\mu mg\, d$

($\vec{f}_k$ and $\vec{d}$ are oppositely directed)

$\Delta KE \quad \Delta K = K_f - K_i = 0 - \frac{1}{2}mV_0^2$

$\therefore -\mu mg\, d = -\frac{1}{2}mV_0^2$

$d = \dfrac{\frac{1}{2}mV_0^2}{\mu mg} = \dfrac{\frac{1}{2}V_0^2}{\mu g} = \dfrac{40.5\ m^2/s^2}{1.96\ m/s^2} = 21\ m$

(b) Since $\vec{N} \perp \vec{d}$, $W = \vec{N} \cdot \vec{d} = 0$.

Since $m\vec{g} \perp \vec{d}$, $W = m\vec{g} \cdot \vec{d} = 0$ ("$\perp$" means perpendicular)

(c) $\vec{f} = m\vec{a}$, $a = \dfrac{f}{m} = \dfrac{\mu mg}{m} = \mu g = 2.0\ m/s^2$

Since $\vec{f}$ is directed to the left, so is $\vec{a}$. The block decelerates.

(d) $P = \dfrac{dW}{dt} = \vec{f_k} \cdot \vec{v}$

Note that since velocity $\vec{v}$ is not constant but decreases, so does Power P.

$P_{AV.} = \left(\vec{f_k} \cdot \vec{v}\right)_{AV.} = \mu_k \, mg \, v_{AV} = \mu_k \, mg \, \dfrac{v+v_0}{2} = \mu_k \, mg \, \dfrac{v_0}{2}$

$\therefore P_{AV.} = 22 \, \dfrac{J}{s} = 22 \, W$

{since a = constant we may write $v_{AV.} = \dfrac{v+v_0}{2}$}

{since v, the final velocity, is zero.}

Also can obtain this result as follows:

$P_{AV.} = \dfrac{\Delta W}{\Delta t} = \dfrac{(\mu mg)(d)}{\Delta t}$,

where Δt is the total time interval for the work done, ΔW.

We need the time interval Δt

$v = v_0 - a\Delta t$

Since $\vec{a}$ is directed to the left (see part (c))

$\therefore \Delta t = \dfrac{v_0}{a}$; $v = 0$ since Δt represents the total transit time.

$P_{AV} = \dfrac{(\mu mg)d}{v_0/a} = \dfrac{(0.20)(2.5)(9.8)(21)}{9.0/2.0} = 23 \, W$,

in agreement with the previous result.

(The small discrepancy comes from rounding of d to two significant figures in part (a).)

5-17

A 5.0 kg block is acted upon by a force parallel to the x-axis and described by the equation:

$$F_x = (4\ N/m^2)\ x^2 - (3\ N).$$

The block moves along the x-axis from x=1.0 m where its velocity is +5.0 m/s to x=2.0 m. Calculate the speed of the block at x=2.0 m if the force whose equation is given above is the only force acting on the block.

****

$$\tfrac{1}{2} m v_{final}^2 = \tfrac{1}{2} m v_0^2 + Work_{1 \to 2}$$

The force F_x is positive between $x=1$ and $x=2$, thus the angle between force and displacement is zero hence:

$$W_{1 \to 2} = \int_1^2 (4x^2 - 3)\, dx = \left(\tfrac{4x^3}{3} - 3x \right) \Big|_1^2 = 6.33\ \text{Joules}$$

$$\tfrac{1}{2} m v_f^2 = \tfrac{1}{2} m v_0^2 + W_{1 \to 2} = \tfrac{1}{2}(5\,kg)(5\,m/s)^2 + 6.33\ J$$

$$= 68.8\ J$$

Thus $v_f = \sqrt{\dfrac{2(68.8)}{5}} = \underline{5.25\ m/s}$

GRAVITATIONAL POTENTIAL ENERGY

5-18

A satellite of mass m revolves around a planet, mass M. The distance r between the satellite and the planet is so large that both objects can be treated as point masses. Suppose the force of gravity between these two bodies is $F = HMm/r^3$, where H is a constant. Derive an expression for the acceleration due to gravity at the position of the satellite. Calculate the change in gravitational potential energy if the satellite is moved a small distance h further from the planet. Give this answer to lowest order in h.

****

Newton's 2nd law: $F = ma \Rightarrow \dfrac{HMm}{r^3} = mg$

$$\therefore g = \dfrac{HM}{r^3}$$

Gravitational potential energy: $U = -\dfrac{HMm}{2r^2} + \text{constant}$

Change in grav. pot. energy $\Delta U = U(r+h) - U(r)$

$$\therefore \Delta U = -\dfrac{HMm}{2}\left[\dfrac{1}{(r+h)^2} - \dfrac{1}{r^2}\right] = \dfrac{HMm}{2r^2}\left[1 - \left(1 + \dfrac{h}{r}\right)^{-2}\right]$$

$$\therefore \Delta U = \dfrac{HMm}{2r^2}\left[1 - \left(1 - 2\dfrac{h}{r} + \cdots\right)\right] = \dfrac{HMmh}{r^3} + \cdots$$

Contrast these results with the usual expressions:

$$g_{usual} = \dfrac{GM}{r^2} \quad \text{and} \quad \Delta U_{usual} \approx \dfrac{GMmh}{r^2} = mgh_{usual}$$

5-19

Consider a binary planetary system consisting of two planets of equal mass, M, and equal radius, R. (They revolve about their common center of mass.) Assume that their center to center distance is 4R and ignore their motion. Then determine an equation, in terms of M and R, for the escape velocity of a projectile shot from the surface of one planet along the line joining the planetary centers but away from both planets.

****

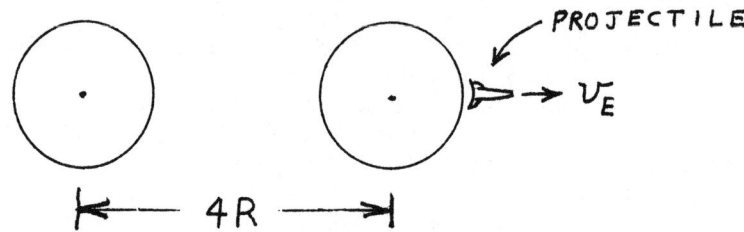

SINCE $E_i = E_f$ (INITIAL = FINAL) AND $E_f = 0$, ONE HAS $E_i = 0$.

OR $K_i + P_i = 0$ (INITIAL KINETIC + INITIAL POTENTIAL E)

SO $\frac{1}{2} m v_E^2 - \frac{GmM}{R} - \frac{GmM}{5R} = 0$

OR $v_E^2 = \frac{2GM}{R}\left(1 + \frac{1}{5}\right) = \frac{12GM}{5R}$ AND $v_E = 2\sqrt{\frac{3GM}{5R}}$.

5-20

A small sphere of mass m is fastened to a weightless string of length 0.5 m to form a pendulum. The pendulum is swinging so as to make a maximum angle of 60° with the vertical.
(a) What is the velocity of the sphere when it passes through the vertical position?
(b) What is the instantaneous acceleration when the pendulum is at its maximum deflection?

****

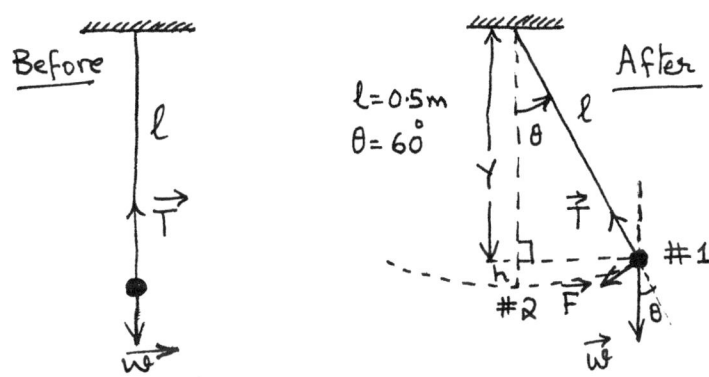

(b) Consider position #1. We have,
$w \cos \theta = T \Rightarrow T = mg \cos 60° = \dfrac{mg}{2}$

Unbalanced force $\vec{F}$ is given by,
$F = w \sin \theta = mg \sin 60° = \dfrac{\sqrt{3}}{2} mg$

$\Rightarrow a \equiv$ instantaneous acceleration $= \dfrac{F}{m} = \dfrac{\sqrt{3}}{2} g$
$ = 0.866 (9.8 \dfrac{m}{\sec^2}) \simeq 8.49 \dfrac{m}{\sec^2}$

(a) At position #1 the sphere has a PE = mgh. When it reaches position #2, the sphere has only KE. If v is the speed of the sphere at position #2 then,
$\dfrac{1}{2} m v^2 = mgh \Rightarrow v^2 = 2gh \Rightarrow v = \sqrt{2gh}$

But, $h = \ell - Y = \ell - \ell \cos \theta = \ell (1 - \cos 60°) = 0.5 m (1 - \dfrac{1}{2}) = 0.25 m$

$\Rightarrow v = \left[2 (9.8 \dfrac{m}{\sec^2}) 0.25 m \right]^{1/2} \simeq 2.21 \dfrac{m}{\sec}$

5-21

An elevator starts from rest and is pulled upward with a constant acceleration of 4 m/sec². The mass of the elevator is 2000 Kg.
 (i) Find tension in the supporting cable.
 (ii) What is the velocity of the elevator after it has risen 15m?
 (iii) Calculate the kinetic energy of the elevator 3 seconds after it starts.
 (iv) Calculate the potential energy of the elevator 3 seconds after it starts.
 (v) What horsepower is required when the speed of the elevator is 8 meters per second?

****

$W = mg$, $m = 2\times10^3$ Kg, $a = 4 \frac{m}{sec^2}$

$\quad\hookrightarrow = 2\times10^3 (Kg) \, 9.8 \left(\frac{m}{sec^2}\right)$

$\Rightarrow W = 1.96 \times 10^4$ N

$F = ma$
$\hookrightarrow = 2\times10^3 \times 4$
$\Rightarrow F = 8\times10^3$ N

(i) $\vec{F} = \vec{T} + \vec{W} \Rightarrow F = T - W$

$\Rightarrow T = F + W = 8\times10^3 (N) + 19.6\times10^3 (N)$
$\quad\hookrightarrow = \underline{\underline{27.6\ N}}$

(ii) $V_f^2 = \cancel{V_i^2}^0 + 2ad = 2\left(4\,\frac{m}{sec^2}\right) 15m = 120\,\frac{m^2}{sec^2}$

$\Rightarrow \underline{\underline{V_f \simeq 10.95\,\frac{m}{sec}}}$

(iii) $V_f' = \cancel{V_i}^0 + at = \left(4\,\frac{m}{sec^2}\right) 3\,sec = 12\,\frac{m}{sec}$

$\Rightarrow KE = \tfrac{1}{2} m V_f'^2 = \tfrac{1}{2}(2\times10^3\,Kg)\,144\left(\frac{m^2}{sec^2}\right) = \underline{\underline{1.44\times10^5\ \text{Joules}}}$

(iv) $Y = \cancel{V_i t}^0 + \tfrac{1}{2}at^2 = \tfrac{1}{2}\left(4\,\frac{m}{sec^2}\right) 9\,sec^2 = 18\,m$

$\Rightarrow PE = mgY = 2\times10^3\,Kg\,(9.8\,\frac{m}{sec^2})\,18m = \underline{\underline{3.53\times10^5\ \text{Joules}}}$

(v) $P \equiv \text{instantaneous power} = T \times V = 27.6(N)\,8\,\frac{m}{sec} \simeq 2.21\times10^5\,\frac{J}{sec}$

$\hookrightarrow = 2.21\times10^5\,\frac{J}{sec}\left(\frac{1Hp}{746\,\frac{J}{sec}}\right) = \frac{2210}{746}\times10^3\,hp \simeq \underline{\underline{296\ Hp}}$

CONSERVATION OF ENERGY

5-22

SUPERMAN (100 Kg) FLIES TO GREAT HEIGHTS BY JUMPING. TO JUMP, HE CROUCHES ON THE GROUND WITH HIS CENTER OF MASS .30 METERS ABOVE THE GROUND. HE JUMPS UP AND LEAVES THE GROUND VERTICALLY WITH HIS CENTER OF MASS .70 METERS ABOVE GROUND. HE REACHES A HEIGHT OF 50 Km. (A) FIND VELOCITY AS HE LEAVES THE GROUND. (B) FIND FORCE EXERTED BY THE GROUND ON HIM (ASSUME CONSTANT).

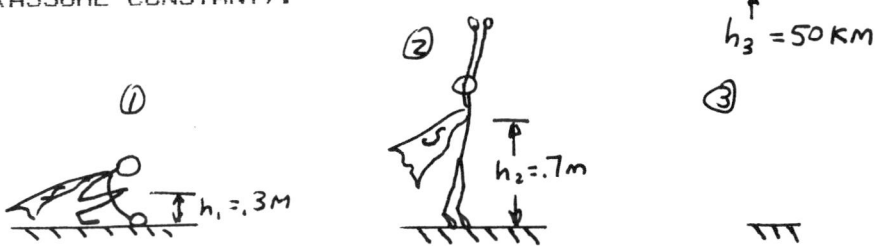

****

A) Since once he leaves the ground, there are no forces other than gravity, Superman's energy is conserved. Therefore, his kinetic energy as he leaves the ground (plus a tiny bit of potential energy) must equal his potential energy at the highest altitude (his velocity = zero there)

1) $\frac{1}{2} m v_1^2 + m g h_1 = \frac{1}{2} m \cancel{v_3^2}^0 + m g h_3$

Solving directly for initial velocity,

$$\frac{1}{2} v_1^2 + g(.7M) = g(50 KM)$$

$$\therefore v_1 = \sqrt{2 \times 9.8 \, m/s^2 \times (50E3M - .7M)}$$

$$= 990 \, m/s$$

B) Since his center of mass location changes due to his own efforts, internal work is done. He pushes against the ground as he "springs" to his take off postion, that is.. going from a crouch to a straight up shape. The external force of the ground on his feet times the change in vertical height of his center of mass equals the internal work that he does.

Using the energy equation, applied work equals the change in potential energy plus the change in kinetic energy..

2) $W_{12} = KE_2 - KE_1 + U_2 - U_1$

$F(h_2 - h_1) = \frac{1}{2} m v_1^2 + m(h_2 - h_1)g$

Solving for the force

$$F = \frac{\frac{1}{2} \times 100 \text{ kg} \left[(990 \text{ m/s})^2 + .4 \text{ m} \cdot 9.8 \text{ m/s}^2 \right]}{.4 \text{ m}}$$

$= 1.22 \text{ E8 N}$

5-23

A projectile is launched with an initial velocity at an unspecified angle above the horizontal. What is the magnitude of its velocity when it reaches a point Y meters below the initial level? Neglect air friction and assume that the gravitational field is constant.

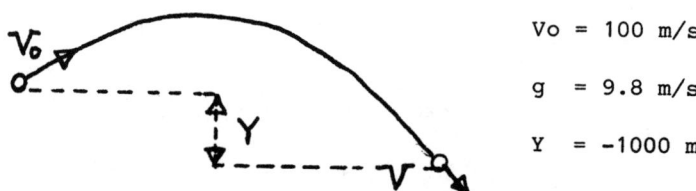

V_o = 100 m/s

g = 9.8 m/s^2

Y = -1000 m

***

ENERGY BEFORE = ENERGY AFTER

$\frac{1}{2} m V_o^2 + mg(0) = \frac{1}{2} m V^2 + mg(Y)$

$V = \sqrt{V_o^2 - 2gY}$, $V = \sqrt{100^2 - 2(9.8)(-1000)} = 172 \frac{m}{s}$

5-24

The 4 kg block slides down a rough surface and has speeds of 3 m/s at point A and 8 m/s at point B. As it slides from A to B, calculate:
(a) the change of its potential energy;
(b) the change of its kinetic energy;
(c) the work of friction.

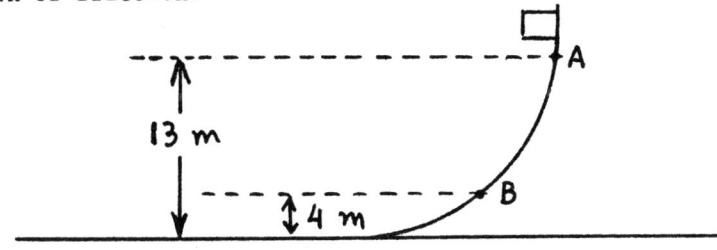

****

(a) $\Delta PE = mg(h_f - h_o) = 4(9.8)(4-13) = \underline{\underline{-352.8 \text{ J}}}$

(b) $\Delta KE = \frac{1}{2} m (v_f^2 - v_o^2) = \frac{1}{2}(4)(8^2 - 3^2) = \underline{\underline{110 \text{ J}}}$

(c) Work-Energy Equation: $W = E_f - E_o$ ∴

$W = PE_f + KE_f - PE_o - KE_o = (PE_f - PE_o) + (KE_f - KE_o)$

$= \Delta PE + \Delta KE = -352.8 + 110 = \underline{\underline{-242.8 \text{ J}}}$

5-25

A 0.4 kg stone is thrown upwards and towards the north. On its return, the stone strikes the ground with a velocity of 20 m/s at an angle of 30° downward from the vertical. Neglect air friction and take ground level as zero potential energy. Gravitational acceleration is 9.8 m/s².

(a) During flight, what was the total mechanical energy of the stone?
(b) At its highest point, what was the velocity (size, direction) of the stone?
(c) How far above ground was the stone at its highest point?

* *

(a) With no friction, the total mechanical energy, E, of the stone doesn't change: $\Delta E = 0$, where
$E = K$ (kinetic energy) $+ U$ (potential energy)
$\quad = \frac{1}{2}mv^2 + mgy$
At impact ($y=0$, $v = 20$ m/s, $m = 0.4$ kg)
$\quad E = \frac{1}{2}(0.4)(20)^2 + 0 = 80$ J (constant during flight)

(b) At the peak height, $v_y = 0$, so $\vec{v}$ is just v_x, to the North. But v_x(top) $= v_x$(ground) because v_x also doesn't change, so
v(top) $= v_x$(ground) $= 20 \sin 30° = 10$ m/s North.

(c) At the peak, $K = \frac{1}{2}mv^2$(top) $= \frac{1}{2}(0.4)(10)^2 = 20$ J
$\quad U = mgY = 0.4 \cdot 9.8 Y = 3.92 Y$
$\quad E = K + U = 20 + 3.92 Y = 80$ (from part "a")
$\therefore Y = \frac{80 - 20}{3.92} = 15.3$ m.

{ Since the problem doesn't state that energy conservation must be used, Y may also be found from the kinematics of free fall. The stone must fall from $v_y = 0$ until v_y becomes $20 \cos 30° = 17.32$ m/s, or for time $= (17.32)/9.8 = 1.767$ s. During that time it falls
$\quad\quad Y = \frac{1}{2}at^2 = \frac{1}{2} \cdot 9.8 \cdot (1.767)^2 = 15.3$ m. }

5-26

A mass m = 2.5 kg is released, after which it slides down a frictionless plane inclined 60° to the horizontal.

(a) Using energy methods find its speed when it has moved along the incline a distance d = 3.5 meters.

(b) Now assume a coefficient of kinetic friction μ_k = 0.20 for the motion. What is now the speed of the block after it has slid the distance d = 3.5(m) along the incline. (HINT: First find the work done by the friction force.)

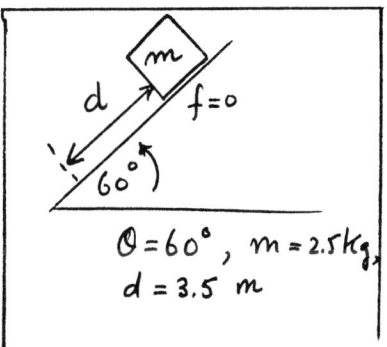

$\theta = 60°$, $m = 2.5$ kg, $d = 3.5$ m

***

(a) $\Delta K = -\Delta U$: the increase in kinetic energy equals the decrease in potential energy

$\frac{1}{2}mv^2 = -mgh$
$\qquad = -mgd\sin 60°$
$v^2 = 2gd\sin 60°$
$v = \sqrt{2gd\sin 60°} = 7.7$ m/s

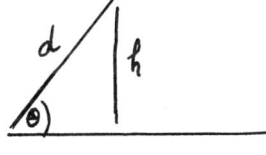

(b) Here, the kinetic energy acquired is reduced from that in part (a) by the amount of the work done by friction force f_k.

$W = f_k d = \mu_k N d = \mu_k mg\cos\theta \, d$
$W = 8.6$ J

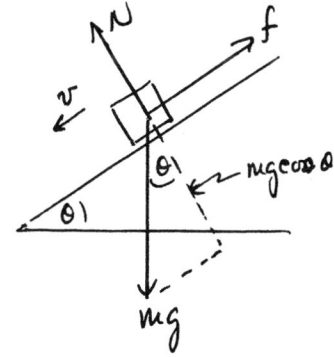

$\therefore \Delta K = \Delta K)_{part(a)} - 8.6 \qquad \Delta K)_{part(a)} = \frac{1}{2}mv^2 = 74$ J

$\therefore \Delta K = 74 - 8.6 = 65$ J $\qquad \therefore \frac{1}{2}mv^2 = 65$
$\qquad\qquad v = \sqrt{2 \times 65/2.5} = 7.2$ m/s

5-27

A 50 kg box is pulled along a horizontal surface by a force of 200 N applied at an angle of 60° above the horizontal. The coefficient of friction of the floor is 0.30. The initial velocity of the box is 5 m/s, and the box is pulled a distance of 10 m. (sin 60° = .866, cos 60° = .500)
a) How much work is done by the applied force?
b) How much work is done by friction?
c) Determine the final velocity of the box by work-energy methods.

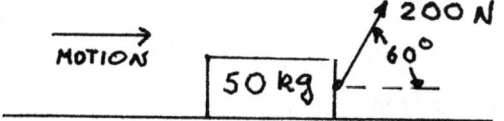

****

a) $W_A = F_A s \cos(60°) = 200(10)(0.500) = 1000$ J

b) To find friction force, need normal force

Free body diagram:

$\Sigma F_{vert} = 0:$ $N - 490 + 200 \sin 60° = 0$

$N = 317$ N

$f = \mu_K N = 95.0$ N

$W_f = -95.0 \times 10 = -950$ J
  ↳ note negative sign

c) $W_A + W_f = \Delta(KE) + \Delta(PE)$

$\Delta(PE) = 0$ $\Delta KE = \frac{1}{2} 50(v_f^2 - 25)$

Thus: $50 = 25(v_f^2 - 25)$

$v_f = \sqrt{\frac{625 + 50}{25}} = 5.20$ m/s

5-28

A 3 kg pendulum bob, suspended by a massless string from point C, is pulled up to point A and released from rest. Find the following quantities when it first reaches point B: (a) the difference between the elevations of points A and B (b) the speed of the bob (c) its centripetal acceleration (d) the tension in the string (e) the tangential acceleration of the bob.

****

(a) Elevation of point A above lowest point (D) of swing is $\overline{ED} = h_1$
$$\overline{ED} = \overline{CD} - \overline{CE} = 1.2 - 1.2\cos 60° = 0.6 \text{ m}$$
$$\therefore h_1 = 0.6 \text{ m}$$

Similarly, elevation of point B above point D (not shown) is $h_2 = 1.2 - 1.2\cos 30° = .161 \text{ m}$

Then $h_1 - h_2 = .600 - .161 = .439 \text{ m}$

(b) $K_1 + U_1 = K_2 + U_2$ Take point B as zero potential energy.
Then $K_1 = 0$ $U_2 = 0$ $U_1 = K_2$
$$mg(h_1 - h_2) = \tfrac{1}{2}mv^2$$
$$V = \sqrt{2g(h_1-h_2)} = \sqrt{2(9.8)(.439)} = 2.93 \text{ m/sec}$$

(c) $a_c = \dfrac{v^2}{R} = \dfrac{(2.93)^2}{1.2} = 7.15 \text{ m/sec}^2$

(d) $\Sigma F = ma$
In radial direction:
$$T - mg\cos 30° = mv^2/R$$
$$T = 3(7.15) + 3(9.8)\cos 30°$$
$$T = 46.9 \text{ N}$$

(e) In tangential direction: $mg\sin 30° = ma_T$
$$a_T = g\sin 30° = 4.9 \text{ m/sec}^2$$

5-29

An 80 kg skier starting from rest at point A on a ski slope, 10 meters above the horizontal, skis down the hill through point B onto a rough horizontal plane. A constant force of kinetic friction of 50 N acts on the skis along the plane which brings the skier to rest at point C. How far does the skier travel from B to C? Assume the slope is frictionless.

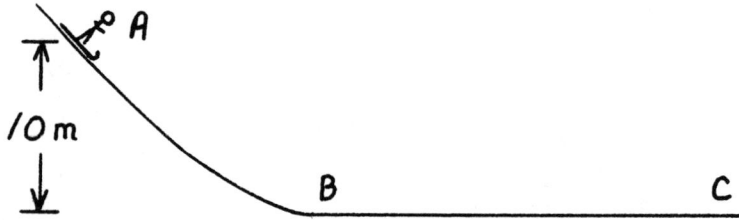

***

Let x = distance BC
E_A = mechanical energy at A
E_C = " " at C

Generalized Work-Energy Th^m $W = \Delta E = E_C - E_A$

where W = work done by friction = $-f \cdot x$
$E_C = 0$
$E_A = mgh$

$\Rightarrow W = E_C - E_A$
$-f \cdot x = 0 - mgh$ $x = \dfrac{mgh}{f} = \dfrac{80(9.8)10}{50} = 157 m$

5-30

The figure shows the track followed by a roller coaster, with the height at each point indicated at the left. The table below is of the kinetic energy, K, and the gravitational potential energy, U. Three entries in the table are given; your job is to fill in the rest of the entries. There is no friction in the problem from A up to D, but when the coaster reaches D the brakes are applied so that a constant frictional force brings it to a stop at point F.

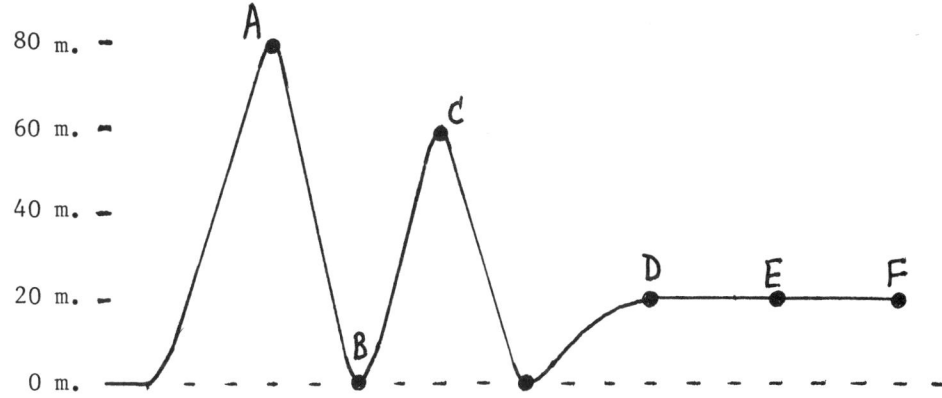

	A	B	C	D	E	F
K	20 kiloJ.					
U	160 kiloJ.	zero				

****

First, the entire "U" row can be filled in from $U = mgh = (2000 \text{ J/m})h$, where I've fitted the given data at A and B.

Next, $(K+U)$ is conserved for A, B, C, and D, $\therefore K = 180 \text{ kJ} - U$.

Finally, from D to F, kinetic energy decreases linearly to zero because of the constant frictional force.

	A	B	C	D	E	F
K	20	180	60	140	70	0
U	160	0	120	40	40	40

5-31

Suppose a particle of mass m is constrained to move only in the X-direction and that the X-position of the particle as a function of time is given by:

$$x(t) = A \cos^2(\omega t)$$

where A and ω are constants.

(a) Derive expressions for the X-components of the velocity and acceleration of the particle as functions of time.

(b) Show that the X-component of the applied force on the particle can be written as a function of x:

$$F(x) = -2m\omega^2 (2x-A)$$

(c) Are any dissipative forces present in this system? Explain your answer.

****

(a) $v_x(t) = \dfrac{dx}{dt} = -2A\omega \cos\omega t \sin\omega t$
$\phantom{v_x(t) = \dfrac{dx}{dt}} = -A\omega \sin 2\omega t$

$a_x(t) = \dfrac{dv_x}{dt} = -2A\omega^2 \cos 2\omega t$

(b) $F_x = ma_x = -2mA\omega^2 \cos 2\omega t$

$\cos 2\omega t = 2\cos^2 \omega t - 1$
$ = 2\left(\dfrac{x(t)}{A}\right) - 1$

$\Rightarrow F_x(x) = -2m\omega^2 [2x - A]$

(c) SINCE F IS ONLY A FUNCTION OF X, THEN A POTENTIAL ENERGY FUNCTION:

$$U(x) = -\int^x F(x)\,dx$$

EXISTS. IF U EXISTS, THEN THE FORCE IS <u>CONSERVATIVE</u> AND THERE CAN BE <u>NO</u> DISSIPATION PRESENT!

5-32

A 2-kg block slides in the frictionless trough as shown. $\overline{AB}$ and $\overline{DE}$ are vertical, and $\overline{BCE}$ is a semicircular segment of radius 4 meters. The block has an initial velocity of 0.4 m/s UPWARD at point A, which is 6 meters above point C. Find the maximum height above point C (at the bottom of the trough) that the block will attain (on the opposite side of the trough), and the force exerted by the block on the trough at point C.

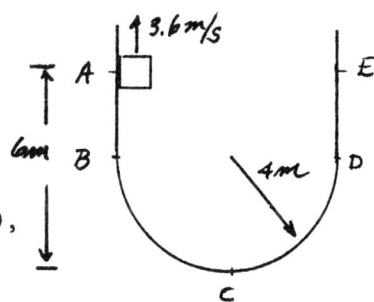

***

Choose point C, the bottom of the trough, as the zero point for measuring potential energy. The total energy of the block at point A is the sum of its kinetic energy ($K = \frac{1}{2}mv^2$) and its gravitational potential energy ($P = mgh$)

$$U = \tfrac{1}{2}mv^2 + mgh$$
$$= \tfrac{1}{2} \cdot 2\,kg \cdot (3.6\,m/s)^2 + 2\,kg \cdot 9.8\,m/s^2 \cdot 6\,m$$
$$= 131\,J$$

When the block reaches its maximum height, all this energy will be in the form of gravitational potential energy $\Rightarrow mgh_{max} = 131\,J \Rightarrow \boxed{h_{max} = 6.66\,m}$

At the bottom of the trough, the block's energy is entirely kinetic $\Rightarrow \tfrac{1}{2}mv^2 = 131\,J \Rightarrow v^2 = 131\,m^2/s^2$

The force exerted on the trough by the block is equal to the force exerted on the block by the trough (Newton's third law) (N in the diagram); this force is NOT just the weight mg! Why not? The block is moving with a speed v in a circle of radius R $\Rightarrow$ it has a centripetal acceleration

$$a_c = v^2/R \Rightarrow \sum \vec{F} = N - mg = ma_c = mv^2/R$$
$$\Rightarrow N = mg + mv^2/R = 2(9.8) + 2(131)/4$$
$$\boxed{N = 84.9\,N}$$

5-33

Masses M and m are fastened to the opposite ends of a string which runs over a pulley wheel. The pulley is restrained from turning and then released. (Treat the string as massless and inextensible, and the pulley wheel as massless and moving without friction.)

(a) Using energy methods find the speed of the masses when m has fallen by 0.5 (m). Take m = 1.5 kg and M = 4.0 kg.

(b) Now taking m to lie on a frictionless plane inclined 30° above the horizontal, with M hanging freely as before, find the speed of the masses when the pulley is released and M has fallen by 0.5 meter.

****

(a) $\Delta K = -\Delta U$, The increase in kinetic energy of the system equals its decrease in potential energy (No work done by dissipative forces in this problem).

$K_0 = \frac{1}{2}(m+M)v_0^2 = 0$; $v_0 = 0$

$K = \frac{1}{2}(m+M)v^2$

$\Delta K = K - K_0 = \frac{1}{2}(m+M)v^2$

$\Delta U = mgd - Mgd$ (m has an increase in its potential energy while M's decreases)

$\therefore \frac{1}{2}(m+M)v^2 = (M-m)gd$

$v^2 = \frac{2(M-m)}{M+m}gd$

$v = 2.1 \text{ m/s}$

(b) When M falls by distance d, m moves up incline by same amount and therefore has a vertical rise of $d \sin 30°$.

$\Delta K = -\Delta U$

$\Delta K = \frac{1}{2}(m+M)v^2$

$\Delta U = mg\, d\sin 30° - Mgd$

or $\Delta K = -\Delta U$ gives in the present case

$$\tfrac{1}{2}(m+M)v^2 = Mgd - mgd \sin 30°$$

$$v^2 = \frac{2(M - m \sin 30°)}{m+M} gd$$

$$v = 2.4 \text{ m/s}$$

5-34

A block of mass m slides down a curved frictionless track and up an incline. The incline has coefficient of sliding friction μ. Calculate the height above the horizontal at which the block comes to rest. Use work-energy methods.

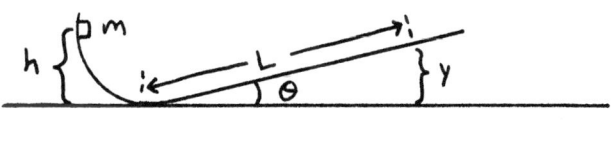

***

$$mgh = \tfrac{1}{2}mv^2_{bottom} = mgy + fL$$

$$f = \mu N = \mu W \cos\theta = \mu mg \cos\theta$$

$$mgh = mgL\sin\theta + \mu mgL\cos\theta$$

$$h = L\sin\theta + \mu L\cos\theta$$

$$L = \frac{h}{\sin\theta + \mu\cos\theta} \quad \text{or}$$

$$y = L\sin\theta = \frac{h\sin\theta}{\sin\theta + \mu\cos\theta} = \frac{h}{1 + \mu\cot\theta}$$

5-35

How much energy is expended in putting a 10,000 kg space ship into a stable circular orbit whose period is 72 hours around the earth?

***

Using conservation of Energy, we know that the total orbital energy (E_f) must equal the total initial Energy (E_i). This E_f has two contributions, one from the gravitational potential energy (U_f) the other from its Kinetic Energy (KE_f). The initial lift-off Energy has a contribution from gravitational potential energy (U_i) and from the required energy (RE) to achieve our desired orbit. Therefore:

$$E_f = E_i$$

$$KE_f + U_f = RE + U_i$$

$$KE_f = \tfrac{1}{2} M v^2 = \tfrac{1}{2} M \left(\frac{2\pi R}{T}\right)^2 \text{ for a circular orbit of radius } R \text{ and period } T$$

From Kepler's 3rd Law:

$$T^2 = \left(\frac{4\pi^2}{GM_e}\right) R^3 \;\rightarrow\; R = \left(\frac{GM_e T^2}{4\pi^2}\right)^{1/3}$$

$$R = \left[\frac{(6.67\times 10^{-11})(6\times 10^{24})(6.72\times 10^{10})}{39.48}\right]^{1/3}$$

$$R = 8.8 \times 10^7 \text{ m}$$

$$KE_f = \tfrac{1}{2}(10{,}000)\left(\frac{2\pi(8.8\times 10^7)}{2.59\times 10^5}\right)^2 = 2.27\times 10^{10} \text{ J}$$

$$U_f = -\frac{GM_e m}{R} = -\frac{(6.67\times 10^{-11})(6\times 10^{24})(10{,}000)}{8.8\times 10^7}$$

$$= -4.55\times 10^{10} \text{ J}$$

$$U_i = -\frac{GM_e m}{R_e} = -\frac{(6.67\times 10^{-11})(6\times 10^{24})(10{,}000)}{6.4\times 10^6}$$

$$= -6.25\times 10^{11} \text{ J}$$

$$\therefore RE = KE_f + U_f - U_i = \underline{\underline{6.03\times 10^{11} \text{ J}}}$$

ELASTIC POTENTIAL ENERGY

5-36

A massless, frictionless spring of force constant 400 N/m is lying on a horizontal surface, with one end fixed to a wall. A 3.00 kg block, which has a coefficient of kinetic friction of 0.300 with the surface, is placed against the spring. The block is then pushed into the spring, compressing it a distance of 0.150 m from equilibrium. The block is then released from rest. Find its speed as it passes through the position where it first touched the spring (the equilibrium position of the spring).

****

Easiest approach: $\begin{pmatrix} \text{Net work done} \\ \text{on the block} \end{pmatrix} = \begin{pmatrix} \text{Change in block's} \\ \text{kinetic energy} \end{pmatrix}$

$$W_{spring} + \underbrace{W_{friction}}_{<0} = \tfrac{1}{2}mv^2 - \underbrace{\tfrac{1}{2}mv_0^2}_{=0}$$

Free-body diagram:

$\uparrow N = mg$
$f = \mu N \leftarrow \square \rightarrow F_{spring} = -kx$
$\downarrow mg$

$$\tfrac{1}{2}kx^2 - \mu mg x = \tfrac{1}{2}mv^2$$

$$\tfrac{1}{2}(400 \tfrac{N}{m})(.150 m)^2 - (.300)(3.00 kg)(9.81 \tfrac{m}{sec^2})(.150 m)$$

$$= \tfrac{1}{2}(3.00 kg) v^2$$

$$\underbrace{(4.50 \tfrac{kg \cdot m^2}{sec^2})}_{\text{Joule}} - (1.32 \tfrac{kg \cdot m^2}{sec^2}) = (1.50 kg) v^2$$

$$v = \underline{\underline{1.46 \tfrac{m}{sec}}}$$

5-37

A mass is dropped from a certain height above the top of a light spring. When it contacts the top of the spring it is latched there. Find the maximum compression and the maximum stretch of the spring. Neglect heat and sound energies.

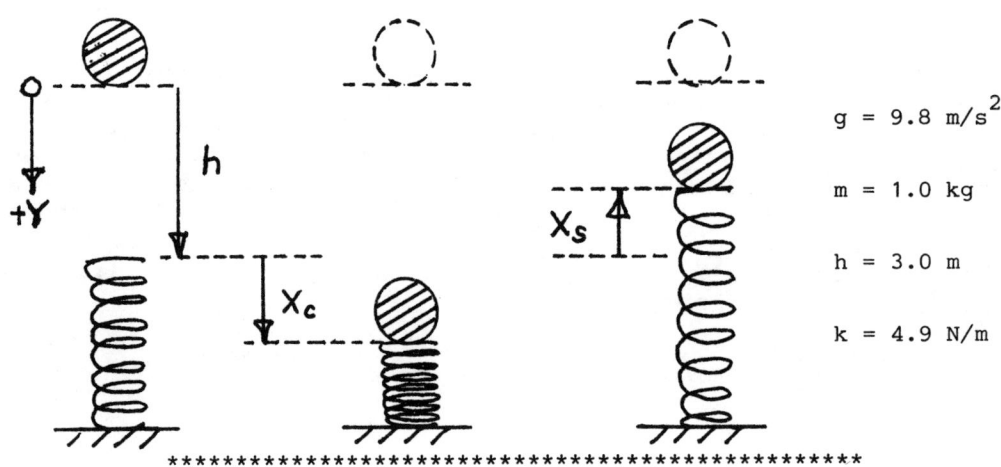

$g = 9.8 \text{ m/s}^2$

$m = 1.0 \text{ kg}$

$h = 3.0 \text{ m}$

$k = 4.9 \text{ N/m}$

***

$\underline{E_{BEFORE}} = \underline{E_{AFTER}}$

$0 = -mg(x+h) + \tfrac{1}{2}kx^2,$

$x^2 - \dfrac{2mg}{k}x - \dfrac{2mgh}{k} = 0, \qquad x^2 - 4x - 12 = 0$

$x = \dfrac{-(-4) \pm \sqrt{(-4)^2 - 4(1)(-12)}}{2(1)}, \qquad \begin{array}{l} X_c = +6 \text{ m} \\ X_s = -2 \text{ m} \end{array}$

ENERGY CHECK:

COMPRESSION: $-(1\text{Kg})(9.8\tfrac{m}{s^2})(6m+3m) + \tfrac{1}{2}(4.9\tfrac{N}{m})(6m)^2$

$= -88.2 \text{ J} + 88.2 \text{ J} = 0$

STRETCHING: $-(1\text{Kg})(9.8\tfrac{m}{s^2})(-2m+3m) + \tfrac{1}{2}(4.9\tfrac{N}{m})(-2m)^2$

$= -9.8 \text{ J} + 9.8 \text{ J} = 0$

5-38

A mass of 2 kg is released from rest on a frictionless incline 10 m from a spring of spring constant k=40N/m. Use g=10 m/s². (a) How fast is the mass moving when it strikes the spring? (b) How far will the spring compress before the mass stops?

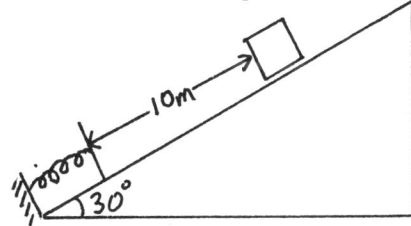

****

(a) Take the top of the uncompressed spring as the zero level of potential energy. Since no work is done by any force other than gravity, mechanical energy is conserved.

$$K_{initial} + U_{initial} = K_{final} + U_{final}$$
$$0 + mgh = \tfrac{1}{2}mv^2 + 0$$
$$v = \sqrt{2gh} = \sqrt{2(10)(10\sin 30°)} = 10 \text{ m/sec}$$

(b) Let the spring compress a distance x, and take the top of the compressed spring as the zero level of potential energy. The mass is initially at a height $(10+x)\sin 30°$ above this level.

$$K_i + U_i = K_f + U_f$$
$$0 + mgh = 0 + \tfrac{1}{2}Kx^2$$
$$2(10)[(10+x)\sin 30°] = \tfrac{1}{2}(40)x^2$$
$$2x^2 - x - 10 = 0$$
$$(2x-5)(x+2) = 0 \qquad x = 2.5 \text{ m}$$

5-39

A spring of spring constant, k=800 N/m, is set in a vertical position with a weightless horizontal platform attached to the top so the spring is not compressed. A 100 kg mass is held just above the platform and released. As the mass accelerates, the spring compresses. This soon causes the mass to deaccelerate. How far is the spring compressed when the velocity of the mass is slowed to zero by the spring? Assume that there are not dissipative or friction forces acting.

****

THE INITIAL AND FINAL GEOMETRIES ARE:

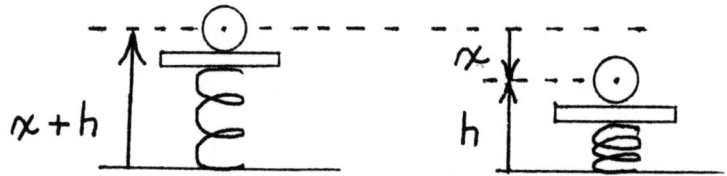

ASSUME THAT THE FALLING MASS COMPRESSES THE SPRING A DISTANCE x AND THE MASS THEN HAS A HEIGHT h. SETTING INITIAL AND FINAL POTENTIAL ENERGIES EQUAL ONE HAS $E_f = E_i$ OR

$$\tfrac{1}{2} k x^2 + mgh = mg(h+x) \quad \text{OR}$$

$$x = \frac{2mg}{k} = \frac{2 \times 100 \text{ kg} \times 9.8 \text{ m s}^{-2}}{800 \text{ N/m}} = 2.45 \text{ m}.$$

NOTE: THIS IS NOT AN EQUILIBRIUM POSITION SINCE THERE IS AN UPWARD FORCE $= 2mg$. THE RESULTING MOTION IS OSCILLATORY.

5-40

A light spring of constant k = 75 N/m has an equilibrium length of 1 m. The spring is compressed to a length of 0.5 m and a mass of 2 kg is placed on its free end on a frictionless slope which makes an angle of 40° with respect to the horizontal. The spring is then released. (a) If the mass is NOT attached to the spring, how far up the slope will the mass move before coming to rest? (b) If the mass IS attached to the spring, how far up the slope will the mass move before coming to rest?

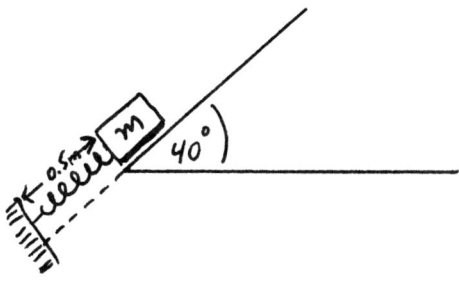

****

(a) By conservation of energy, the final energy (E_f) which will be made up of only gravitational potential energy (PE_g) must equal the initial energy (E_i) which had only a contribution from the spring potential energy (PE_s).

$$E_f = E_i$$
$$PE_g = PE_s$$
$$mg(\ell \sin\theta) = \tfrac{1}{2}kx^2$$
$$\ell = \frac{\tfrac{1}{2}kx^2}{mg \sin 40°} = \frac{\tfrac{1}{2}(75)(0.5)^2}{(2)(9.8)\sin 40°} = 0.744 \, m$$

(b) This time, the final energy has a contribution from the extension of the spring (PE_s').

$$E_f = E_i$$
$$PE_g + PE_s' = PE_s$$
$$mg(\ell_2 \sin\theta) + \tfrac{1}{2}kx_2^2 = \tfrac{1}{2}kx^2$$
$$x_2 = \ell_2 - 0.5$$
$$x_2^2 = \ell_2^2 - \ell_2 + 0.25$$

In this case we get a quadratic equation to solve:

$$\tfrac{1}{2}k\ell_2^2 + (mg\sin 40° - \tfrac{1}{2}k)\ell_2 = 0$$

giving $\ell_2 = 0.664 \, m$

5-41

On a horizontal table, a 2-kg block is attached by a horizontal spring to a fixed point. The work which must be done on the spring to extend it 0.25 m beyond its unstretched length is 4 joules.

(a) If the block is held at rest so that the spring is stretched 0.25 m, what force will the spring exert on it?
(b) If the block is released from its position in part (a), and if the table is frictionless, what will be the block's speed when the spring returns to its unstretched position?
(c) Actually, when the spring becomes unstretched after release from the position in part (a), the speed of the block is 1.6 m/s. What constant friction force did the table exert on the block?

* *

Let x be the horizontal displacement of the 2 kg block from the equilibrium position (where spring is unstretched).

The force the spring exerts on the block is proportional but opposite to x:

(1) $\qquad F = -kx \qquad k \equiv$ force constant of spring

To extend such a spring requires an applied force $F^a = -F = +kx$ and the work done by F^a is

(2) $\qquad \text{Work} = \int_0^x F^a\, dx = \int_0^x kx\, dx = \frac{1}{2}kx^2$

and this is also the potential energy of the stretched spring.

(a) In Eq.(2), we are given Work = 4 J for $x = 0.25$ m:
$\qquad 4 = \frac{1}{2} k (.25)^2 \quad$ so $\quad k = 128$ N/m

From Eq. (1)
$\qquad$ Force exerted by spring $= -128$ N/m $(0.25\,m) = -32$ N.,
the minus sign indicating direction.

(b) With no friction, the total mechanical energy, E, is constant.
$\qquad E =$ Kinetic energy of block + potential energy of spring
$\qquad\quad = \frac{1}{2}mv^2 + \frac{1}{2}kx^2$
At A: $v = 0$, $x = 0.25$ m $\qquad E_A = 0 + \frac{1}{2}(128)(.25)^2 = 4$ J
At B: $x = 0 \qquad E_B = \frac{1}{2}mv_B^2 + 0 = \frac{1}{2}\cdot 2\,kg \cdot v_B^2$
But $E_A = E_B$ so
$\qquad v_B^2 = 4 \Rightarrow v_B = 2$ m/s.

(c) In this case $E_B = \frac{1}{2} m v_B^2 = \frac{1}{2} \cdot 2 \cdot (1.6)^2 =$ only 2.56 J because the friction force, F_k, has transformed mechanical energy into thermal energy (heat). By the work-energy theorem
  Work done on block by $F_R = \Delta E = E_B - E_A = 2.56 - 4.0$
    $= -1.44$ J

Since F_R is constant, the work it does is also
  $F_R \times$ (displacement) $\times \cos\{$angle between F_R and displacement$\}$

The displacement AB is 0.25 m and F_R is opposite to it so the angle is π rad:
  Work by $F_R = F_R(0.25) \cos\pi = -0.25 F_R = -1.44$ J.
    or $F_R = 5.76$ N

{If μ_R is needed it is now available. Because the table is horizontal, the normal force N it exerts on the block is the weight of the block, $mg = 2 \times 9.8 = 19.6$ N.
  $\mu_R = F_R / N = \frac{5.76}{19.6} = 0.29$ }

5-42

A spring is placed in a vertical tube as shown. A mass of 0.4 kg is placed on the spring and pushed down until the spring is compressed a distance of 0.1 m below its uncompressed length. When the mass is released, how high above the point from which it is released will the mass travel? The mass is not fastened to the spring. Neglect friction. The spring constant is 200 nt/m.

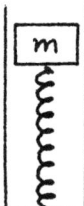

****

The elastic potential energy stored in the compressed spring reappears as the gravitational potential energy of the mass at its highest position, that is,

$$\tfrac{1}{2}(200)(0.1)^2 = (0.4)(9.8)\, y$$

$$\therefore\ y = \frac{1}{(0.4)(9.8)} \cong 0.255\ (m) \leftarrow$$

(The mass of spring is neglected.)

5-43

An 8 kg mass is pushed against a horizontal spring (force constant 3200 N/m) attached to a vertical wall until the spring is compressed 0.25 m less than its natural length. The mass is then released from rest and the spring is allowed to expand, pushing the mass into motion along a frictionless surface shown in the diagram. The spring and mass are not connected, so the mass moves freely independent of the spring after they lose contact.

a) Find the kinetic energy of the mass just after the spring has reached its natural length.

b) What would be the speed of the mass when it reaches point A in the diagram, which is a vertical distance of 2 m below the level the spring is on?

c) What would be the vertical height h of point B, where the speed of the mass instantaneously becomes zero? Note h is measured from point A.

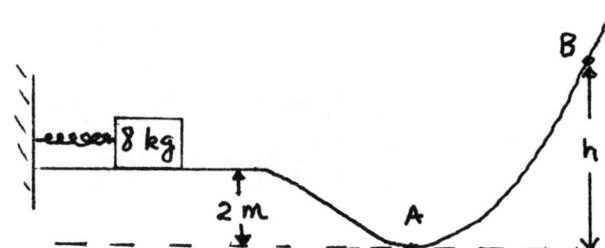

****

a) $E = KE +$ Gravitational $PE +$ Spring PE

$$= \tfrac{1}{2}mv^2 + mgh' + \tfrac{1}{2}kx^2$$

where h' is the height above reference point. Let reference point be A.

x is the compression of the spring

There is no friction so Total Energy E is conserved.

At Release: $\tfrac{1}{2}mv^2 = 0 \qquad mgh' = 8(9.8)(2) = 157 J$

$$\tfrac{1}{2}kx^2 = \tfrac{1}{2}3200(.25)^2 = 100 J$$

Thus $\qquad E = 257 J$

For spring at natural length $x = 0$

This first occurs while $h' = 2m$

Thus $\qquad 257 = \tfrac{1}{2} 8(v^2) + 157 + 0$

$$v^2 = 25 \; m^2/s^2$$

$$v = 5 \; m/s$$

b) At Point A: $\qquad h' = 0$

$$257 = \tfrac{1}{2}mv^2 + 0$$

$$v = \sqrt{\tfrac{257}{4}} = 8.02 \; m/s$$

c) At point B: $\qquad h' = h \qquad v = 0$ (it's stopped)

$$257 = (8)(9.8)h$$

$$h = 3.28 \; m$$

5-44

A block of mass .5 kg is placed against a spring of spring constant 1000 N/m attached to one end of an inclined plane inclined 53° as shown below. The spring is compressed .2 m and the block is released from rest. Use conservation of energy methods to find the velocity of the block after it has slid 2 m up the plane if the coefficient of friction between the block and the plane is .3.

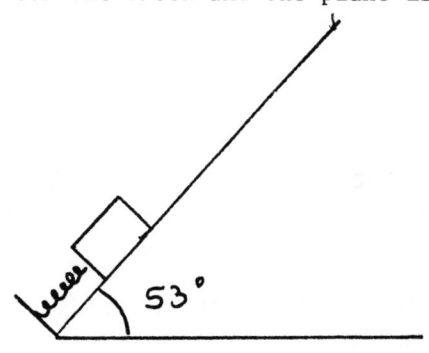

***

$W_f = \Delta K + \Delta U = K_2 + U_{2s} + U_{2g} - (K_1 + U_{1s} + U_{1g})$

$W_f = -(\mu mg \cos 53°)(2m)$

$= -(.3)(.5 kg)(9.8 m/s^2)(.6)(2m) = -1.76 J$

$U_{1s} = \frac{1}{2} k x^2$

$U_{1s} = \frac{1}{2}(1000 N/m)(.2m)^2 = 20 J \qquad U_{2s} = 0$

$U_{2g} - U_{1g} = mg(y_2 - y_1) = (.5 kg)(9.8 m/s^2)(2m \sin 53°)$
$\qquad\qquad = 7.84 J$

$-1.76 J = \frac{1}{2} m v_2^2 + 7.84 J - 20 J$

$20 J - 7.84 J - 1.76 J = \frac{1}{2}(.5 kg) v_2^2$

$v_2 = 6.4 \text{ m/s}$

6
IMPULSE AND MOMENTUM

IMPULSE AND MOMENTUM

== 6-1

A 1700kg car travels at 25m-s^{-1} and collides head on with a concrete overpass support. The support does not move and the car comes to rest against the support. At the instant of collision, the center of mass of the car is 2.5m from the concrete surface. After the collision, the center of mass of the car is 1.2m from the concrete surface.
a) Find the average stopping force exerted on the car by the concrete.
b) Find the time interval of the stopping of the car.

****

a.)
$d = 2.5m - 1.2m = 1.3m$

$\text{Work} = Fd = \Delta(\text{Kinetic Energy}) = \frac{1}{2}mv^2$

$d = 1.3m \Rightarrow F = \frac{\frac{1}{2} \times 1700 \times (25)^2}{1.3} = 4.08 \times 10^5 N$

b.) $\text{Impulse} = Ft = \Delta p = m\Delta v = 1700 \times 25$

$\therefore t = \frac{\Delta p}{F} = \frac{1700 \times 25}{4.08 \times 10^5} = 0.104 s$

6-2

A 1.0 Kg SOCCER BALL IS DROPPED TO THE GROUND FROM A HEIGHT OF 5.0 METERS. IT BOUNCES STRAIGHT UP TO A HEIGHT OF 5.0 METERS. (A) FIND THE IMPULSE GIVEN TO THE BALL BY THE GROUND. (B) IF THE BALL WAS IN CONTACT WITH THE GROUND FOR .05 SEC, WHAT WAS THE AVERAGE FORCE EXERTED BY THE GROUND ON THE BALL? (C) DRAW THE VECTOR DIAGRAM SHOWING THE MOMENTUM VECTORS JUST BEFORE AND JUST AFTER THE BALL HITS THE GROUND, AND THE IMPULSE VECTOR.

****

A) The impulse applied to the ball equals the change in momentum vector of the ball before and after it hits the ground.

Value of momentum vector just before it hits the ground

1) $\bar{P}_1 = m\bar{V}_1 = 1\,kg \cdot \bar{V}_1$

Value of momentum vector just after it hits the ground

2) $\bar{P}_2 = m\bar{V}_2 = 1\,kg \cdot \bar{V}_2$

In both cases have to know the velocity. When not in contact with the ground, energy of the ball is constant. Therefore, kinetic energy at ground level equals potential energy at the 5 meter height.

3) $\frac{1}{2}mv^2 = mgh$

Solving for velocity

$$V = \sqrt{2gh} = \sqrt{2 \cdot 9.8\,m/s^2 \cdot 5m} = 9.9\,m/s$$

Impulse/momentum equation ($\bar{j}$ is unit vector in $+y$ direction)

4) $\bar{I} = \bar{P}_2 - \bar{P}_1 = m(\bar{V}_2 - \bar{V}_1) = mV(\bar{j} - (-\bar{j})) = 2mV\bar{j}$

Plugging in the values of velocity and solving for impulse

5) $\bar{I} = 2 \cdot 1\,kg \cdot 9.9\,m/s = 19.8\,kg\,m/s\,\bar{j}$

B) If force is constant, it is:

$$\bar{I} = \int \bar{F}\,dt = \bar{F}\Delta t$$

$$\therefore \bar{F} = \bar{I}/\Delta t = \frac{19.8\,kg\,m/s}{.05\,s} \cdot \frac{N}{kg\,m/s^2}\,\bar{j} = 396\,N\,\bar{j}$$

C) Remember, impulse and momentum are vectors, and have direction and magnitude associated with them.

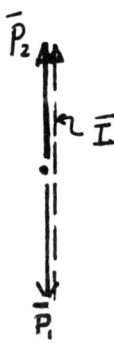

6-3

A 20.0 g bullet initially traveling at 600 m/s penetrates 9.00 cm into an "infinitely" massive wooden block.
(a) What average force does the block exert on the bullet?
(b) How long does it take the block to bring the bullet to rest?

****

(a) Bullet's initial kinetic energy = Work done in stopping it.

$$\tfrac{1}{2}mv^2 = Fx$$

$$F = \frac{mv^2}{2x} = \frac{(.0200\,kg)(600\,m/sec)^2}{(2)(.0900\,m)} = \underline{\underline{4.00 \times 10^4\,N}}$$

(b) Bullet's initial momentum = Impulse given to it to stop it.

$$mv = Ft$$

$$t = \frac{mv}{F} = \frac{(.0200\,kg)(600\,m/sec)}{(4.00 \times 10^4\,N)} = \underline{\underline{3.00 \times 10^{-4}\,\text{second}}}$$

6-4

A ball of mass 80 g is dropped from 4 meters onto a hard floor. If it rebounds to a height of only 3 meters, calculate the impulse exerted on the ball by the floor.

****

before impact

$h_o = 4m$, E_o at top, E_f at bottom with V_b downward

impact

Impulse $= \vec{J}$ (upward)

after impact

$h_f = 3m$, E_f at top, E_o at bottom with V_a upward

$E_o = E_f$

$mgh_o = \frac{1}{2}mV_b^2$

$V_b = \sqrt{2gh} = \sqrt{2(9.8)(4)}$

$V_b = 8.85 \text{ m/s}$

$E_o = E_f$

$\frac{1}{2}mV_A^2 = mgh_f$

$V_A = \sqrt{2gh_f}$

$V_A = 7.67 \text{ m/s}$

$\Rightarrow \vec{J} = \Delta \vec{p} = \vec{P_a} - \vec{P_b} = m\vec{V_a} - m\vec{V_b}$

$|\vec{J}| = .08(7.67) + .08(8.85)$

$|\vec{J}| = 1.32 \text{ N·s}$ upward

6-5

A steel ball whose mass is 0.1 kg and whose speed is 50 m/sec, strikes a vertical wall at an angle of $30°$ as shown in the figure. If the ball leaves the wall with a speed of 50 m/sec and is in contact with the wall for 0.01 sec, what average force did the ball exert on the wall?

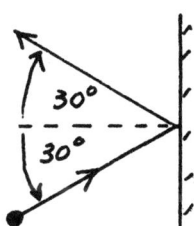

****

From newton's 2nd Law

$$\vec{F} = \frac{d\vec{p}}{dt} \Rightarrow \Delta\vec{p} = \int_{t_1}^{t_2} \vec{F}\, dt$$

For an average Force

$$\Delta\vec{p} = \vec{F}_{ave} \int_0^{\Delta t} dt = \vec{F}_{ave}\, \Delta t$$

The change in momentum of the ball is

$$\Delta\vec{p} = \vec{P}_f - \vec{P}_i = 2 \times (-.1 \times 50 \times \cos 30)\, \vec{i}$$

$$= -8.66\, \vec{i} \ \ kg\, m/sec$$

∴ Force acting on the ball

$$\vec{F}_{ave} = \frac{\Delta\vec{p}}{\Delta t} = -\frac{8.66\, \vec{i}}{.01} = -866\, \vec{i}\ N$$

The force on the wall, by newton's 3rd Law is

$$\vec{F}_{reaction} = -\vec{F}_{ave} = +866\, \vec{i}\ N$$

6-6

If an airplane going 600 mph strikes a six-inch bird which weighs ½ pound, estimate the force of the bird on the airplane.

****

Describing problem:

$V_a = 600$ mph
$\quad = 880$ ft/s (before and after)

Assume that the initial velocity of the bird is essentially zero and that the final velocity of the bird is that of the airplane which remains essentially unchanged.

$\ell_b = .5$ ft $\quad W_b = .5$ lb. $\quad V_{ib} = 0 \quad V_{fb} = 880$ ft/s

Setting up eqs.

Use the impulse-momentum theorem

$$F \Delta t = \Delta p \quad \Rightarrow \quad F = \frac{m_b \Delta v_b}{\Delta t} = \frac{m_b V_a}{\Delta t}$$

Assume that Δt is equivalent to the time that it takes for the plane to travel the length of the bird.

$$\Rightarrow \Delta t = \ell_b / V_a$$

So $\quad F = \frac{m V_a}{\ell_b / V_a} = \frac{m_b V_a^2}{\ell_b} = \frac{W_b V_a^2}{g \ell_b}$

Substituting numbers:

$$F = \frac{(.5)(880)^2}{(32)(.5)} = \underline{\underline{2.4 \times 10^4 \text{ lb.}}} \approx 11 \text{ tons !!}$$

6-7

Two particles with masses 2.0 kg and 3.0 kg have velocities $-12.0\,\hat{i} + 6.0\,\hat{j}$ m/sec and $12.0\,\hat{i} - 4.0\,\hat{j}$ m/sec, respectively.

a) Calculate the linear momentum of each particle.
b) Calculate the total linear momentum of the system.
c) Calculate the velocity of the center of mass of the system.
d) Calculate the acceleration of the center of mass.

****

a) $\bar{P}_A = M_A \bar{V}_A = 2.0\,kg\,(-12.0\,\hat{i} + 6.0\,\hat{j}\text{ m/s})$

$\qquad = -24.\,\hat{i} + 12.\,\hat{j}\text{ kg m/s}$

$\bar{P}_B = M_B \bar{V}_B = 3.0\,kg\,(12.0\,\hat{i} - 4.0\,\hat{j}\text{ m/s})$

$\qquad = 36.\,\hat{i} - 12.\,\hat{j}\text{ kg m/s}$

b) $\bar{P}_{TOT} = \bar{P}_A + \bar{P}_B = (-24.\,\hat{i} + 12.\,\hat{j}) + (36.\,\hat{i} - 12.\,\hat{j})\,\tfrac{kg}{m/s}$

$\qquad = 12.\,\hat{i}\text{ kg m/s}$

c) $\bar{V}_{CM} = \dfrac{\bar{P}_{TOT}}{M_{TOT}} = \dfrac{12.\,\hat{i}\text{ kg m/s}}{2.0 + 3.0\,kg} = 2.4\,\hat{i}\text{ m/s}$

d) $\bar{a}_{CM} = \dfrac{d\bar{V}_{CM}}{dt} = \dfrac{d}{dt}(2.4\,\hat{i}\text{ m/s}) = 0$

Or $\bar{F}_{ext} = M_{TOT}\,\bar{a}_{CM} = \dfrac{d\bar{P}_{TOT}}{dt} = \dfrac{d}{dt}(12.\,\hat{i}\text{ kg m/s})$

$\qquad = 0$

6-8

A 2 kg mass is thrown from a window 5 m above a scaffolding plank, so that it lands on the plank and stops. The initial velocity of the mass is 10 m/s at an angle of 27° with the downward vertical. Calculate the impulse delivered to the plank, and state its direction. If the collision with the plank takes 1 msec. calculate the average force exerted on the plank during the impact.

****

Take the x-axis horizontal and the y-axis vertically downward. The velocity on impact has components:

$$v_x = v_0 \sin 27° = 10 \sin 27° = 4.54 \text{ m/s}$$

$$v_y^2 = v_{oy}^2 + 2gh = (10 \cos 27°)^2 + 2(9.8)(5) \; (\tfrac{m}{s})^2$$

$$\therefore v_y = 13.3 \text{ m/s}$$

Impulse $\vec{I} = \Delta \vec{p}$

$I_x = m v_x$
$\therefore I_x = (2 \text{ kg})(4.54 \text{ m/s})$
$I_y = m v_y$
$\therefore I_y = (2 \text{ kg})(13.3 \text{ m/s})$

$$\therefore \vec{I} = (9.08, 26.6) \text{ N-s}$$

$$|\vec{I}| = \sqrt{I_x^2 + I_y^2} = \sqrt{(9.08)^2 + (26.6)^2} = 28.1 \text{ N-s}$$

at angle $\theta = \tan^{-1} \dfrac{I_x}{I_y} = \tan^{-1} \dfrac{9.08}{26.6} = 18.8°$

with the downward vertical.

$$|\langle \vec{F} \rangle| = \dfrac{I}{\Delta t} = \dfrac{28.1 \text{ N-s}}{10^{-3} \text{ s}} = 28100 \text{ N}$$

CONSERVATION OF LINEAR MOMENTUM

6-9

A projectile of mass 0.10 kg originally travelling at a speed of 10 m/s collides with a 10.0 kg stationary target. Upon collision the original projectile and a 0.50 kg piece of the target continue together with a speed of 5.0 m/s along the original direction of the projectile. Using the collision approximation (no outside forces), calculate the velocity of the other (9.5 kg) part of the original target after the collision.

$v_0 = 10$ m/s
$m = 0.10$ kg
$V = 0$
$M = 10.0$ kg
BEFORE

$V' = ?$
$v' = 5.0$ m/s
$m + \Delta m = 0.6$ kg
AFTER

***

The linear momentum is conserved in our approximation, hence:

$$p_{before} = p_{after}$$

$$mv_0 + MV = (m + \Delta m)v' + (M - \Delta m)V'$$

$$(0.1 \text{ kg})(10 \text{ m/s}) = (0.6 \text{ kg})(5 \text{ m/s}) + (9.5 \text{ kg})V'$$

$$V' = \frac{-2.0 \text{ kg m/s}}{9.5 \text{ kg}} = \underline{\underline{-0.21 \text{ m/s}}}$$

(Yes, the target rebounds to the <u>left</u>!)

6-10

A bullet of mass 10^{-2} kg traveling at 600 m/s strikes a wood block of mass .5 kg from below as shown. The bullet emerges from the block traveling at 100 m/s. (a) What is the velocity of the block right after the bullet leaves it? (b) How high does the block rise? (c) If the bullet was in contact with the block for .0001 s, what was the average force exerted by the bullet on the block?

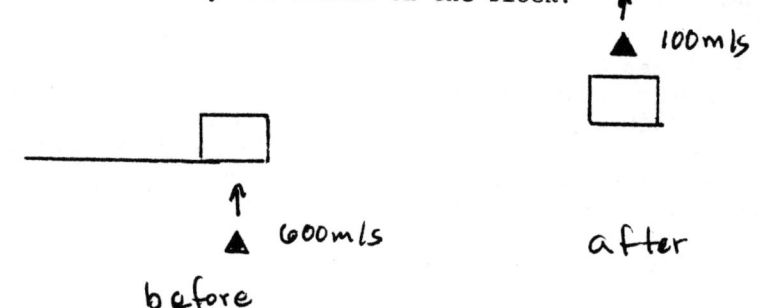

****

a) $P_{y\;initial} = P_{y\;final}$

$$(10^{-2} kg)(600 m/s) + 0 = (10^{-2} kg)(100 m/s) + (.5 kg) v_{block}$$

$$v_{block} = \frac{(6\;kg\text{-}m/s) - 1\;kg\text{-}m/s}{.5\;kg} = 10 m/s$$

b) $\Delta K + \Delta U = 0$

$$\Delta K = 0 - \tfrac{1}{2} m v^2 = -\Delta U = -mgh$$

$$h = \frac{v^2}{2g} = \frac{(10 m/s)^2}{2(9.8 m/s^2)} = 5.1 m$$

c) Impulse $= F_{AV} \Delta t = \Delta p$

$$F_{AV} = \frac{(.5 kg)(10 m/s) - 0}{.0001 s} = 5 \times 10^4 N$$

6-11

Consider two particles of masses m_1 and m_2 which interact via a <u>repulsive</u> force such that the force on m_2 due to m_1 has the form:

$$\vec{F}_{21}(r) = \frac{k}{r^3}\hat{r}$$

where r is the distance between them and $\hat{r}$ is the unit vector pointing from m_1 to m_2.

(a) Show that the potential energy of the system as a function of r can be written in the form:

$$U(r) = \frac{k}{2r^2}.$$

(b) Suppose that the particles are initially at rest a distance r_0 apart, and at a later time t, they are a distance $2r_0$ apart. If the particles experience no external forces outside of their mutual interaction, what is the <u>total</u> kinetic energy of the two particle system at t?

(c) Find an expression for the speeds v_1 and v_2 of m_1 and m_2 respectively at t.

****

(a) $$U(r) = -\int_{\infty}^{r}\vec{F}(\vec{r})\cdot d\vec{r} = -\int_{\infty}^{r}\frac{k}{r^3}dr$$

$$\boxed{U(r) = \frac{k}{2r^2}}$$

(b) ENERGY IS $\overline{\text{CONSERVED}}$ $\Rightarrow K_i + U_i = K_f + U_f$

BOTH PARTICLES INITIALLY AT REST $\Rightarrow K_i = 0$.

$$U_i = \frac{k}{2r_0^2} \qquad U_f = \frac{k}{8r_0^2} \quad \Rightarrow \quad \boxed{K_f = \frac{3}{8}\frac{k}{r_0^2}}$$

(c) MOMENTUM IS CONSERVED SINCE THERE ARE NO EXTERNAL FORCES $\Rightarrow \vec{P}_f = \vec{P}_i = 0$.

$$\Rightarrow m_1 v_1 = m_2 v_2, \quad v_1 = \frac{m_2 v_2}{m_1}$$

$$K_f = \frac{1}{2}m_1 v_1^2 + \frac{1}{2}m_2 v_2^2 = \left(\frac{m_2}{m_1}+1\right)\left(\frac{1}{2}m_2 v_2^2\right)$$

$$v_2^2 = \frac{2K_f m_1}{m_2(m_1+m_2)} \quad \Rightarrow \quad v_2 = \sqrt{\frac{3k\, m_1}{4m_2(m_1+m_2)r_0^2}}$$

$$v_1 = \frac{m_2}{m_1}v_2 = \sqrt{\frac{3k\, m_2}{4m_1(m_1+m_2)r_0^2}}$$

6-12

A star of mass 2×10^{30} kg moving with a velocity of 2×10^4 m/s collides with a second star of mass 5×10^{30} kg moving with a velocity of 3×10^4 m/s at right angles to the first. If they join together what is their common velocity?

****

Conservation of linear momentum applies. Momentum is a vector quantity, so the x and y components are conserved separately.

x - direction

before = after

$$M_1 V_1 = (M_1 + M_2) V_x$$

$$2 \times 10^{30} \times 2 \times 10^4 = (2+5) \times 10^{30} V_x$$

$$V_x = (4/7) \times 10^4 = 5.7 \times 10^3 \text{ m/s}$$

For the y-direction

$$M_2 V_2 = (M_1 + M_2) V_y$$

$$V_y = (15/7) \times 10^4 = 21.4 \times 10^3 \text{ m/s}$$

so

$$V^2 = V_x^2 + V_y^2 \quad \text{gives} \quad V = 22.1 \times 10^3 \text{ m/s}$$

$$\tan \theta = V_y / V_x = 3.75 \qquad \theta = 75°$$

6-13

A 2000 kg automobile going east on Lomas Street at 60 km·hr^{-1} collides with a 4000 kg truck which is going southward across Yale Street at 20 km·hr^{-1}. If they become coupled on collision, what is the magnitude and direction of their momentum immediately after colliding?

****

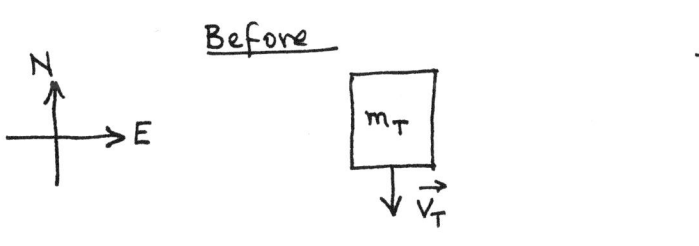

Before   After

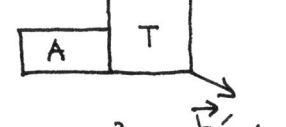

$V_A = 60 \times 10^3 \frac{m}{hr} \left(\frac{1 hr}{3600 sec}\right) = \frac{50}{3} \frac{m}{sec}$, $m_A = 2 \times 10^3$ Kg, $\vec{P'} = (m_A + m_T)\vec{V'}$

$V_T = \frac{1}{3} V_A = \frac{50}{9} \frac{m}{sec}$, $m_T = 4 \times 10^3$ Kg

One can see that the collision is completely inelastic. But linear momentum **must** be conserved.

$\Rightarrow \vec{P_A} + \vec{P_T} = \vec{P'} \Rightarrow \vec{P} = \vec{P'}$

$P_A = m_A V_A = 2 \times 10^3$ Kg $\left(\frac{50}{3} \frac{m}{sec}\right)$
$\quad \hookrightarrow = 3.33 \times 10^4$ Kg m/sec

$P_T = m_T V_T = 4 \times 10^3$ Kg $\left(\frac{50}{9} \frac{m}{sec}\right) \simeq 2.22 \times 10^4$ Kg $\frac{m}{sec}$

$\Rightarrow P = \sqrt{P_A^2 + P_T^2} = \sqrt{(3.33)^2 + (2.22)^2} \times 10^4$ Kg $\frac{m}{sec}$
$\quad \hookrightarrow \simeq 4 \times 10^4$ Kg $\frac{m}{sec}$

$\Rightarrow \underline{P' = 4 \times 10^4 \text{ Kg } \frac{m}{sec}}$

$\theta = \arctan(P_T/P_A) = \arctan\left(\frac{2}{3}\right) \simeq 33.69°$

$\Rightarrow \underline{\theta \simeq 33.69° \text{ South of east}}$

6-14

Two pendulum-balls are pulled aside as shown. When they are released, they collide head on and stick together at the bottom of their swing, and then they rise together. How high will they rise? The masses of the balls are 0.1 kg and 0.2 kg, respectively, and the length of the massless cords are both 0.5 m.

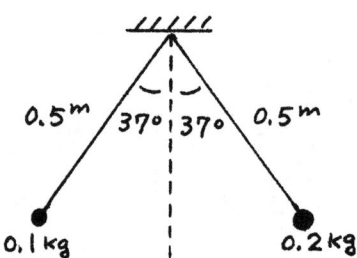

****

The speed of each ball before the impact v is, neglecting air resistance, $v = \sqrt{(2)(9.8)(0.5)(1-\cos 37°)} = 1.4$ m/sec.

Applying the linear momentum conservation principle,

$$(0.1)(1.4) - (0.2)(1.4) = [(0.1)+(0.2)]V,$$

where V is the speed of two balls together after the impact. And solving for V,

$$V = -0.47 \text{ m/sec} \quad (\text{to the left}).$$

If the balls rise to h m above their lowest position, neglecting air resistance again,

$$\tfrac{1}{2}(0.1+0.2)V^2 = (0.1+0.2)(9.8)h$$

$$\therefore h = \frac{V^2}{(2)(9.8)} = \frac{(0.47)^2}{(2)(9.8)} = 1.1 \times 10^{-2} \text{ m} \leftarrow$$

ELASTIC COLLISIONS

6-15

A 3 kg object, sliding on a frictionless surface, makes an elastic collision with an object of unknown mass initially at rest. If the 3 kg object rebounds from the collision with one-fourth its original speed, what is the unknown mass?

****

Before Collision: $[3\,Kg] \to u \quad [m]$

After Collision: $\frac{u}{4} \leftarrow [3\,Kg] \quad [m] \to v$

Energy is conserved

$$\tfrac{1}{2}(3)u^2 = \tfrac{1}{2}(3)\tfrac{u^2}{16} + \tfrac{1}{2}mv^2$$

$$3u^2 - \tfrac{3}{16}u^2 = mv^2$$

$$\tfrac{45}{16}u^2 = mv^2$$

Momentum is conserved

$$3u = mv - 3 \times \tfrac{u}{4}$$

$$3u + \tfrac{3}{4}u = mv$$

$$\tfrac{15}{4}u = mv$$

Substitute for mv

$$\tfrac{45}{16}u^2 = (mv)v = \tfrac{15}{4}uv$$

$$\tfrac{3}{4}u = v$$

$$\Rightarrow \tfrac{15}{4}u = mv = m\left(\tfrac{3}{4}u\right) \quad \Rightarrow \quad m = 5\,Kg$$

6-16

Two gliders are placed on a frictionless air track which defines the x-axis. Glider #1 has mass $m_1 = 0.2$ kg and is initially at $x = -1.2$ m and is moving with a velocity of +1.5 m/sec. Glider #2 has mass $m_2 = 0.4$ kg and is initially at rest at $x = 0$. (a) Initially, where is the center of mass of the system? (b) When glider #1 hits glider #2 an elastic collision occurs. What will be the final velocities of the two gliders?

```
          →
         [M₁]                    [M₂]
  ────────┼──────────────────────┼────────
       x = -1.2 m               x = 0
```

****

(a) $$X_{cm} = \frac{\sum_i m_i x_i}{\sum_i m_i} = \frac{(0.2)(-1.2) + (0.4)(0)}{0.2 + 0.4} = -0.4 \text{ m}$$

(b) For mass M_1, use the formula

$$V_{1f} = \left(\frac{M_1 - M_2}{M_1 + M_2}\right) V_{1i}$$

$$= \left(\frac{0.2 - 0.4}{0.2 + 0.4}\right)(1.5) = \underline{\underline{-0.5 \text{ m/sec}}}$$

For mass M_2, use the formula

$$V_{2f} = \left(\frac{2M_1}{M_1 + M_2}\right) V_{1i}$$

$$= \frac{2(0.2)}{0.2 + 0.4}(1.5) = \underline{\underline{1.0 \text{ m/sec}}}$$

Check the conservation of momentum:

<u>initially</u>
$$M_1 V_1 + M_2 V_2 = (0.2)(+1.5) + (0.4)(0)$$
$$= 0.3 \text{ kgm/sec}$$

<u>finally</u>
$$M_1 V_1 + M_2 V_2 = (0.2)(-0.5) + (0.4)(1.0)$$
$$= 0.3 \text{ kgm/sec} \quad \checkmark$$

6-17

A neutron moving at 600 m/sec collides head-on with a stationary deuteron. Assume that the deuteron has exactly twice the mass of the neutron. If this isolated two-body collision is completely elastic, calculate the speed of the deuteron after the collision. Then find the velocity of the neutron after the collision.

****

Before: neutron (m) moving at $v = 600 \frac{\text{meters}}{\text{sec}}$ toward stationary deuteron (2m).
After: neutron (m) with $v_1' = ?$, deuteron (2m) with $v_2' = ?$

MOMENTUM: $mv + (2m)(0) = mv_1' + (2m)v_2'$

$$v = v_1' + 2v_2'$$

KINETIC ENERGY: $\frac{1}{2}mv^2 + \frac{1}{2}(2m)(0)^2 = \frac{1}{2}mv_1'^2 + \frac{1}{2}(2m)v_2'^2$

$$v^2 = v_1'^2 + 2v_2'^2$$

COMBINING: $v_1'^2 = (v - 2v_2')^2$

$$\cancel{v^2} = \cancel{v^2} - 4vv_2' + 4v_2'^2 + 2v_2'^2$$

$$4vv_2' = 4v_2'^2 + 2v_2'^2 = 6v_2'^2$$

$$4v = 6v_2' \implies v_2' = \tfrac{2}{3}v = \tfrac{2}{3}\left(600 \tfrac{m}{\text{sec}}\right)$$

The speed of the deuteron after the collision,

$$\boxed{v_2' = 400 \tfrac{m}{\text{sec}}}.$$

Then $v_1' = v - 2v_2' = 600 \tfrac{m}{\text{sec}} - 2\left(400 \tfrac{m}{\text{sec}}\right)$

The speed of the neutron after the collision, $|v_1'| = 200 \tfrac{m}{\text{sec}}$ but the velocity of the neutron after the collision, $\boxed{v_1' = -200 \tfrac{m}{\text{sec}}}$, that is, 200 $\tfrac{m}{\text{sec}}$ in the direction opposite to its original velocity.

6-18

Two objects move without friction across a horizontal surface toward each other and undergo a perfectly elastic head-on collision. If the masses and velocities before collision are 4 kg, 6 m/s and 2 kg, 3 m/s, find the velocities of both objects after the collision.

Suppose that the collision had been perfectly inelastic. What would the final velocity be and how much kinetic energy would be lost during the collision?

***

Let's begin by moving into a frame of reference in which computation will be easier. Consider a frame constantly moving to the left at 3 m/s along with the 2 kg object before the collision. As seen from this frame, the 2 kg object is at rest and the 4 kg object moves to the right at 9 m/s. Now, both linear momentum and kinetic energy are conserved in a perfectly elastic collision so, in the new frame

$$P = (4\,kg)(+9\,m/s) = +36\,kg\,m/s \text{ (to right)}$$

and $K.E. = \frac{1}{2}(4)(9)^2 = 162$ Joules

The conservation equations are then (after collision)

$$(*)\; 36 = 4U_4 + 2U_2 \quad ; \quad (**)\; 162 = \frac{1}{2}(4)U_4^2 + \frac{1}{2}(2)U_2^2$$

Solving * for U_2 gives $U_2 = 18 - 2U_4$ and

Plugging this into ** gives

$$162 = 2U_4^2 + (18 - 2U_4)^2 = 2U_4^2 + 324 - 72U_4 + 4U_4^2$$

OR $\quad 6U_4^2 - 72U_4 + 162 = 0$

This can be reduced to $(U_4 - 9)(U_4 - 3) = 0$

and therefore $U_4' = +9$ or $U_4' = +3$.

The first cannot be physically correct if there is a collision. Therefore $U_4' = +3$ m/s after collision

Since $U_2' = 18 - 2U_4'$, $U_2' = +12$ after collision

But these velocities are not in the frame of reference in which the problem was originally stated. Converting back into the original frame by subtracting 3 m/s from each of the velocities above gives $U_4' = 0$ and $U_2' = 9$ m/s

To check that we are correct we should compute the linear momentum and kinetic energy in the original frame before and after collision getting

$$(4kg)(6 \text{ m/s}) + (2kg)(-3 \text{ m/s}) = 18 \text{ kg m/s} = (2kg)(9 \text{ m/s})$$

and

$$\tfrac{1}{2}(4kg)(6)^2 + \tfrac{1}{2}(2)(3)^2 = 81 \text{ Joules} = \tfrac{1}{2}(2kg)(9 \text{ m/s})^2$$

6-19

Consider two identical hard steel balls hung freely as shown in figure A. If ball #1 is pulled away to the left (figure B) and allowed to swing into ball #2 with speed V_1, determine how many balls will swing outward from the right immediatly after the collision. Assume the collision is elastic.

figure A   figure B

***

From cons. of momentum:
$$m\vec{v_1} = m\vec{v_1}' + m\vec{v_2}'$$ (primes denote after collision parameters)

$$\therefore \vec{v_1} = \vec{v_1}' + \vec{v_2}' \qquad (1)$$

From cons. of kinetic energy:
$$\tfrac{1}{2}mv_1^2 = \tfrac{1}{2}mv_1'^2 + \tfrac{1}{2}mv_2'^2$$

$$\therefore v_1^2 = v_1'^2 + v_2'^2 \qquad (2)$$

Squaring (1) yields
$$v_1^2 = v_1'^2 + v_2'^2 + 2\vec{v_1}'\cdot\vec{v_2}' \qquad (3)$$

Subtracting (2) from (3):
$$2\vec{v_1}'\cdot\vec{v_2}' = 0 \quad \text{and since the 'dot'} = 1 \cdot 1$$

either $\vec{v_1}' = 0$ or $\vec{v_2}' = 0$

If $\vec{v_2}' = 0$ then $\vec{v_1}' = \vec{v_1}$ i.e. ball #1 missed ball #2.

If $\vec{v_1}' = 0$ then $\vec{v_2}' = \vec{v_1}$ and only ball #2 moves off to the right.

$\therefore$ In either case only <u>one</u> ball moves off to the right!

INELASTIC COLLISIONS AND THE COEFFICIENT OF RESTITUTION

6-20

During the perfectly inelastic collision between the two bodies shown, an impulse of 40 Newtons-sec to the right is applied by and outside agent. Calculate their velocity after collision.

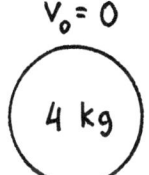
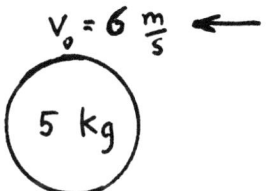

***

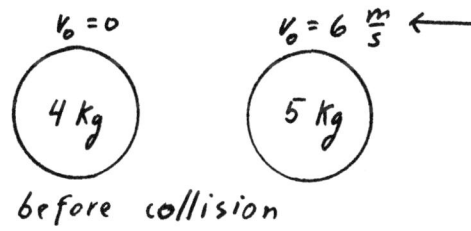
before collision

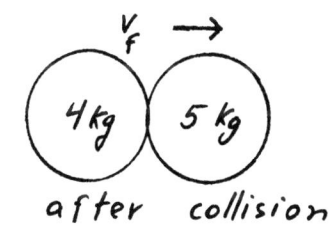
after collision

Impulse - Momentum Equation: $I = \Sigma mv_f - \Sigma mv_0$

$\xrightarrow{+}$ $40 = (4+5)v_f - 5(-6)$

$9v_f = 40 - 30$ and $v_f = 1.11 \text{ m}/_s \rightarrow$

6-21

A 500 gram glob of putty is thrown horizontally at a 4.0 kg block, initially at rest on a level, frictionless floor. The putty sticks to the block, and the combination slides across the floor, impacts a horizontal, frictionless spring of force constant 900 N/meter, and compresses the spring 10 cm. What was the initial speed of the putty?

****

Let v = speed of putty before impact
V = speed of combination after impact

On collision with the spring, energy is conserved.
$$K_1 + U_1 = K_2 + U_2$$
$$\tfrac{1}{2}(4.5 \text{ kg}) V^2 + 0 = 0 + \tfrac{1}{2}(900 \text{ N/m})(.1 \text{ m})^2$$
$$V = \sqrt{2} = 1.414 \text{ m/sec}$$

Now consider the original collision. Momentum is conserved, but kinetic energy is not.
$$mv = (m+M) V$$
$$(.5 \text{ kg}) v = (4.5 \text{ kg})(1.414 \text{ m/sec})$$
$$v = 12.7 \text{ m/sec}$$

Inelastic Collisions And The Coefficient Of Restitution / 215

6-22

When a bullet mass of 20 g strikes a ballistic pendulum of mass 10 kg, the center of gravity of the pendulum is observed to rise a vertical distance of 7 cm. The bullet remains embedded in the pendulum.

(a) Calculate the original velocity of the bullet.
(b) What fraction of the original kinetic energy of the bullet remains as kinetic energy of the system immediately after the collision?

****

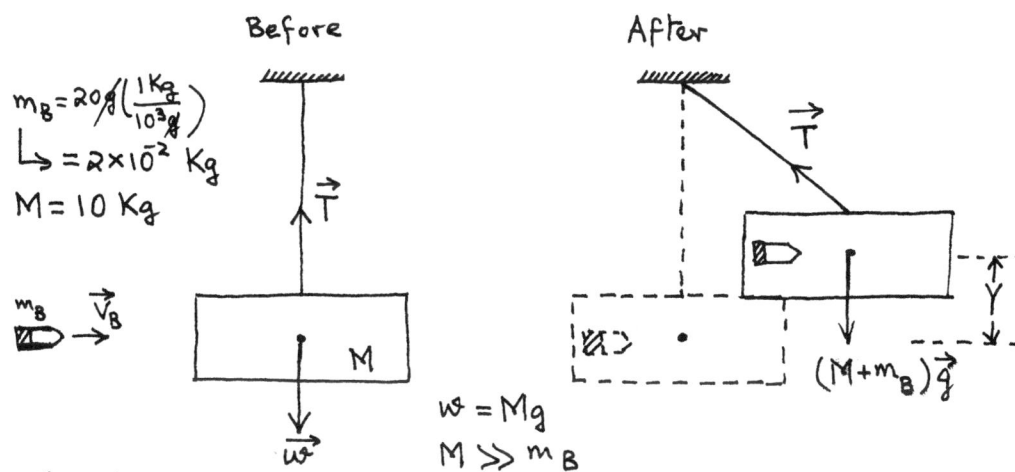

$m_B = 20 g \left(\frac{1 Kg}{10^3 g}\right)$
$\hookrightarrow = 2 \times 10^{-2}$ Kg
$M = 10$ Kg

$W = Mg$
$M \gg m_B$

This is a case of completely inelastic collision.

(a) Immediately after the collision the linear momentum of the system is conserved. Let V' be the velocity of the block after the collision,

$$m_B V_B = (M + m_B) V' \Rightarrow V_B = \left(1 + \frac{M}{m_B}\right) V' \simeq \frac{M}{m_B} V' \quad —(1)$$

Also,

$$\tfrac{1}{2}(M+m_B) V'^2 = (M+m_B) g Y \Rightarrow V' = \sqrt{2gY} \quad —(2)$$

From equations (1) and (2) we get,

$$V_B = \frac{M}{m_B}\sqrt{2gY} = \frac{10\,Kg}{2\times 10^{-2} Kg}\left(2 \times 9.8 \tfrac{m}{sec^2}\right)7\times 10^{-2} m\Big)^{1/2} = 500(1.372)^{1/2}\,\tfrac{m}{sec}$$

$$\Rightarrow V_B \simeq \underline{585.67\,\tfrac{m}{sec}}$$

(b) $\dfrac{T_{system}}{T_{bullet}} = \dfrac{\tfrac{1}{2}(M+m_B)V'^2}{\tfrac{1}{2} m_B V_B^2} \simeq \dfrac{M}{m_B}\left(\dfrac{V'}{V_B}\right)^2 = \dfrac{M}{m_B}\left(\dfrac{m_B}{M}\right)^2 = \dfrac{m_B}{M}$

$\hookrightarrow = \dfrac{2\times 10^{-2} Kg}{10 Kg} = 2\times 10^{-3} = 0.002$

$\hookrightarrow \ll 1$ We would expect this result, since KE is not conserved in an inelastic collision.

6-23

Two objects move across a horizontal frictionless surface directly toward each other and make a partially inelastic head-on collision with each other. If the masses and velocities are: 3 kg, 10 m/s, and 5 kg, 2 m/s, and the coefficient of restitution is 0.11, find:

(a) the velocity of each object after the collision.
(b) the kinetic energy change.
(c) the average force acting on each, if the collision lasts for 0.1 second.

---

a) Let's work in a frame of reference moving with velocity 2 m/s to the left along side of the 5 kg object before the collision. In this frame the initial velocities are

$$U_3 = +12 \text{ m/s} \text{ and } U_5 = 0$$

before collision.
Indicate the velocities after collision by a Prime (v)

$$e = \frac{\text{relative velocity after}}{\text{relative velocity before}} = \frac{|U_3' - U_5'|}{12} = 0.11$$

Hence either * $U_3' - U_5' = 1.32$ or ** $U_5' - U_3' = 1.32$

The equation of conservation of momentum is

*** $(3 \text{ kg})(12 \text{ m/s}) = 36 = 3U_3' + 5U_5'$

Solving this and equation ** simultaneously gives

$36 = 3(U_5' - 1.32) + 5U_5' = 8U_5' - 4 \; ; \; U_5' = 5 \; ; \; U_3' = 3\frac{2}{3}$.

Solving *** and * simultaneously leads to the physically impossible solution $U_5' = 4$ and $U_3' = 5\frac{1}{3}$

Converting back into the original frame by subtracting 2 gives
$U_5' = 3 \text{ m/s}$ and $U_3' = (5/3) \text{ m/s}$

b) Calculating Kinetic energies after collision we get
$$K.E_5' = \tfrac{1}{2}(5kg)(3^{m/s})^2 = 22.5 \text{ Joules}$$
$$K.E_3' = \tfrac{1}{2}(3)(1.67)^2 = 4.2 \text{ Joules}$$
Whereas, the kinetic energy before collision was
$$K.E_5 = \tfrac{1}{2}(5)(2)^2 = 10 \text{ J}$$
$$K.E_3 = \tfrac{1}{2}(3)(10)^2 = 150$$
Therefore the change in Kinetic energy is
$$\Delta K.E = 26.7 - 160 = -133 \text{ Joules}$$
c) The average force exerted by 3 on 5 is
$$\bar{F}_{ave} = \frac{m_3 \Delta v_3}{\Delta t} = (3kg)\left(\frac{1.67 - 10^{\,m/s}}{0.1 s}\right) = -250 \text{ New.}$$
The average force exerted on 5 by 3 is
$$\bar{F}_{ave} = \frac{m_5(\Delta v_5)}{\Delta t} = (5)\left(\frac{3-(-2)}{0.1}\right) = +250 \text{ New.}$$

6-24

A 0.01-kg bullet is fired into a 1.99 kg block. The block is attached to a spring (spring constant 500 N/m), and the block (containing the bullet) slides without friction to the right, compressing the spring 0.4 m. What was the velocity of the bullet just before it struck the block?

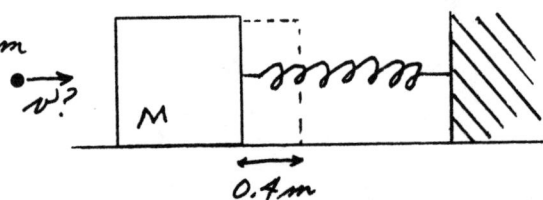

****

Conserve ENERGY after collision; conserve MOMENTUM during collision.

energy of block/bullet/spring system when spring compressed (block stopped) (elastic potential energy) $= \frac{1}{2} k x^2 = \frac{1}{2} \cdot 500 \cdot .4^2 = 40 \text{ J}$

energy of block/bullet/spring just after impact (block moving, spring not compressed) (all kinetic energy) $= \frac{1}{2}(M+m)V^2$

$\Rightarrow \frac{1}{2}(M+m)V^2 = \frac{1}{2}(1.99 \text{ kg} + 0.01 \text{ kg})V^2 = 40 \text{ J}$

$\Rightarrow V = 6.32 \text{ m/s}$

Conserve momentum:

initial momentum $= mv$ (bullet alone) $= (m+M)V$ (bullet imbedded in block)

$\Rightarrow \boxed{v = \frac{(m+M)V}{m} = 1260 \text{ m/s}}$

6-25

A car C of mass 1500 kg is moving east with a constant speed of 12 m/s toward an intersection. A 5000 kg truck T is moving south at a constant speed of 4 m/s toward the same intersection. They collide at the intersection and the wreckage sticks together. Five seconds before the collision the car is 60 m west of the intersection and the truck is 20 m north of the intersection.

a) Determine the coordinates of the center of the mass of the car-truck system five seconds before the collision. Use the positive x-direction east and the positive y-direction north with the origin at the point of impact. Treat the car and the truck as particles.
b) Determine the components of the velocity of the center of mass before the collision.
c) Determine the components of the velocity of the wreckage just after the collision. Give reasoning to support your answer.

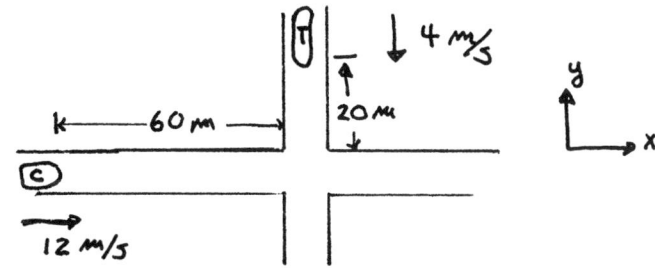

****

a) $X_{CM} = \dfrac{X_c \cdot M_c + X_T \cdot M_T}{M_c + M_T} = \dfrac{-60(1500) + 0}{6500} = -13.8\,m$

$Y_{CM} = \dfrac{Y_c \cdot M_c + Y_T \cdot M_T}{M_c + M_T} = \dfrac{0 + 20(5000)}{6500} = 15.4\,m$

b) $V_{xCM} = \dfrac{V_{xc} \cdot M_c + V_{xT} \cdot M_T}{M_c + M_T} = \dfrac{(12)(1500) + 0}{6500} = 2.77\,m/s$

$V_{yCM} = \dfrac{V_{yc} \cdot M_c + V_{yT} \cdot M_T}{M_c + M_T} = \dfrac{0 - 4(5000)}{6500} = -3.07\,m/s$

c) The Momentum of the Car-Truck system is $\vec{P} = (M_c + M_T)\vec{V}_{CM}$ and is conserved in the collision. The velocity of the wreckage is the same as in (b).

6-26

The diagram below shows a head-on collision between a 1 kg mass with speed 25 m/s and a 2 kg mass with a speed of 10 m/s. After the collision their speeds are called v and V, respectively. The table below gives five different values of v; for each of these five different collisions, find V and the change in kinetic energy, $\Delta K = K_f - K_i$, and fill your results into the table.

Of the five collision cases (labelled A,B,C,D,E), which are elastic? Why? Which are completely inelastic? Why?

BEFORE: [1] →25 m/s [2 KG] →10 m/s   After: [1] →v   [2 KG] →V

	A	B	C	D	E
v	−5 m/s	0	+5 m/s	+10 m/s	+15 m/s
V					
ΔK					

****

Conserve momentum: $(1\,kg)(25\,m/s) + (2\,kg)(10\,m/s) = 1v + 2V$

$$\boxed{22.5\,m/s - \tfrac{1}{2} v = V}$$

Now that we know V for any given v, we can calculate

$$\Delta K = K_f - K_i = \tfrac{1}{2}(2\,kg)V^2 + \tfrac{1}{2}(1\,kg)v^2 - \left[\tfrac{1}{2}(2\,kg)10^2 + \tfrac{1}{2}(1)25^2\right]$$

$$\boxed{\Delta K = V^2 + \tfrac{1}{2} v^2 - 412.5 \text{ JOULES}}$$

We can use these two eqs to fill in V and ΔK in the table:

	A	B	C	D	E
v	−5	0	+5	+10	+15
V	25	22½	20	17½	15
ΔK	225 J.	94 J.	0	−56 J.	−75 J.

Only case C is elastic: only there is $\Delta K = 0$.

Only case E is "completely inelastic": there $v = V = 15\,m/s$.

This problem is designed, in part, to help you see where

the term "completely inelastic" comes from; notice that if you combine the two boxed equations to eliminate V, you get for the energy loss, $K_i - K_f = -\Delta K = 75 - \frac{3}{4}(\nu - 15)^2$ JOULES : that is the equation of a parabola with its <u>maximum at $\nu = V = 15$ m/s</u> : that is, if $\nu = V$ after the collision, then as much kinetic energy has been lost as is possible (while conserving momentum). Thus the term "COMPLETELY INELASTIC".

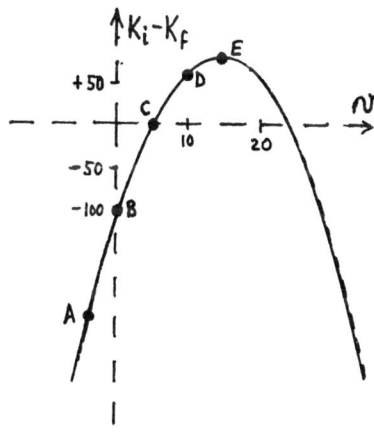

6-27

A mass (m_1 = 3 kg) slides along a frictionless surface with a speed of 3 m/sec. It hits and attaches to a second mass (m_2 = 1 kg) which is connected to a spring of constant k = 40 N/m originally at its equilibrium length. What will be the maximum compression of the spring?

****

First we must use conservation of momentum (p) to determine the speed of the combined masses after the collision:

$$p_f = p_i$$

$$(m_1+m_2)v = m_1v_1 + m_2v_2 = (3)(3) + (1)(0)$$
$$(m_1+m_2)v = 9 \text{ kgm/sec}$$
$$v = 9/(3+1)$$
$$v = 2.25 \text{ m/sec}$$

Now we use conservation of Energy (E) to determine the compression of the spring when the masses come to a halt:

$$E_f = E_i$$

$$\tfrac{1}{2}kx^2 = \tfrac{1}{2}(m_1+m_2)v^2$$

$$x = \sqrt{\frac{m_1+m_2}{k}}\, v = \sqrt{\frac{3+1}{40}}\,(2.25) = \underline{\underline{0.71 \text{ m}}}$$

6-28

Two 2.0 kg masses, A and B, collide. The velocities before the collision were V_A = 15 i and 30 j m/sec and V_B = -10 i + 5.0 j m/sec. After the collision the velocity of particle A is V_{Af} = -5.0 i + 20 j m/sec.

a) Calculate the final velocity of particle B.
b) Calculate how much energy was lost or gained in the collision.
c) If the collision took 1.0 milliseconds, what was the average force of particle A on particle B?

****

a) $m_A \bar{V}_A + m_B \bar{V}_B = m_A \bar{V}_{Af} + m_B \bar{V}_{Bf}$. $M_A = M_B$

$15\hat{i} + 30\hat{j} + (-10\hat{i} + 5.0\hat{j}) = -5.0\hat{i} + 20\hat{j} + \bar{V}_{Bf}$

$\bar{V}_{Bf} = 10.\hat{i} + 15.\hat{j}$ m/s

b) $\frac{1}{2} m_A V_A^2 + \frac{1}{2} m_B V_B^2 = \frac{1}{2} m_A V_{Af}^2 + \frac{1}{2} m_B V_{Bf}^2 - \Delta K.E.$

$\Delta K.E. = \frac{1}{2} m [V_{Af}^2 + V_{Bf}^2] - \frac{1}{2} m [V_A^2 + V_B^2]$

$= \frac{2.0}{2} [(-5)^2 + (20)^2 + (10)^2 + (15)^2] - \frac{2.0}{2} [(15)^2 + (30)^2 + (-10)^2 + (5)^2]$

$= [25 + 400 + 100 + 225] - [225 + 900 + 100 + 25]$

$= 750 - 1250 = -500$ joules .

Thus, 500 joules of mechanical energy was lost during the collision.

c) $\bar{F}_{av} = \frac{\Delta \bar{P}_A}{\Delta t} = \frac{m_A (\bar{V}_{Af} - \bar{V}_A)}{1.0 \times 10^{-3} \text{ sec}}$

$= \frac{(2.0)[(-5.0 - 15)\hat{i} + (20 - 30)\hat{j}]}{1.0 \times 10^{-3}}$ $\frac{kg \, m/sec}{sec}$

$= -4.0 \times 10^4 \hat{i} - 2.0 \times 10^4 \hat{j}$ Newtons

6-29

Assume one-dimensional motion for three isolated blocks, with initial velocities as shown. Blocks A and B undergo an elastic collision; then B and C undergo a completely inelastic collision. Derive an equation for the speed of B and C after they join together.

***

Speed of A, B center of mass: $v_{CM} = v_A \dfrac{m_A}{m_A + m_B}$.

First collision viewed relative to center of mass

$u_A = v_A - v_{CM} = v_A \dfrac{m_B}{m_A + m_B}$, $u_B = 0 - v_{CM} = -v_A \dfrac{m_A}{m_A + m_B}$

Relative to CM, an elastic collision simply reverses velocities, so $u_B' = +v_{CM}$, and $v_B' = u_B' + v_{CM} = v_A \dfrac{2m_A}{m_A + m_B}$ ← speed of B after first collision

(also!)

Second collision conserves linear momentum:

$m_B v_B' = (m_B + m_C) V \;\Rightarrow\; \boxed{V = v_A \dfrac{2 m_A m_B}{(m_A + m_B)(m_B + m_C)}}$

A simple case can be used as a check: If the three masses are equal, $v_B' = v_A$ and $V = \tfrac{1}{2} v_A$.

6-30

A 100 gram projectile collides inelastically with a 2.0 kilogram block, which is initially at rest. The projectile sticks to the block, which is given a speed of 3.0 m/s by the collision. Calculate the initial speed and kinetic energy of the projectile.
How much mechanical energy was lost in the collision?

****

Projectile: $m_1 = 0.10 \text{ kg}$ $v_1 = ?$
Block: $m_2 = 2.0 \text{ kg}$ $v_2 = 3.0 \text{ m/s}$

Use Conservation of Momentum:

$$\vec{P}_{initial} = \vec{P}_{final}$$

$$m_1 v_1 = (m_1 + m_2) v_2$$

$$v_1 = \frac{m_1 + m_2}{m_1} v_2$$

$$v_1 = 63 \text{ m/s}$$

Mechanical energy means kinetic plus potential. In this problem only kinetic energy changes.

$$\Delta E = \Delta K = K_{FINAL} - K_{INITIAL}$$

$$= \tfrac{1}{2}(m_1 + m_2) v_2^2 - \tfrac{1}{2} m_1 v_1^2$$

$$= 9.5 \text{ J} \quad - 200 \text{ J}$$

$$\Delta E = -190 \text{ J}$$

That is, 190 J was lost in the collision.
(Actually, it was not lost, just changed into energy forms like heat, sound, etc.)

6-31

A neutron moving at 600 m/sec collides head-on with a stationary deuteron. Assume that the deuteron has exactly twice the mass of the neutron. If this isolated two-body collision is completely inelastic, calculate the speed of the tritium which emerges.

***

$\overset{n}{\underset{m}{\bigcirc}} \longrightarrow v' = \dfrac{600 \text{ meters}}{\text{sec}}$ $\qquad \underset{2m}{\bigcirc d} \text{ at rest}$

BEFORE COLLISION: Momentum of the two-body system $= mv' + 0$

AFTER COLLISION: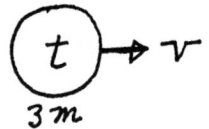
$\underset{3m}{\bigcirc t} \longrightarrow v$

Momentum of the tritium $= (3m)v$

CONSERVATION OF MOMENTUM $\Rightarrow$

$(3m)v = mv' \Rightarrow 3v = v' \Rightarrow$

$v = \dfrac{v'}{3} = \dfrac{600 \text{ m/sec}}{3} \Rightarrow \boxed{v = 200 \dfrac{\text{meters}}{\text{sec}}}$

6-32

A 2.0 kg mass is propelled at a stationary 8.0 kg mass. After a head-on, partially elastic collision characterized by a coefficient of restitution of 0.5, the 8.0 kg mass recoils into a light horizontal spring attached to a wall. (See the figure below.) If the original speed of the 2.0 kg mass was 5.0 m/s and the 8.0 kg mass compressed the spring by 0.2 m at the maximum, what is the stiffness constant of the spring?

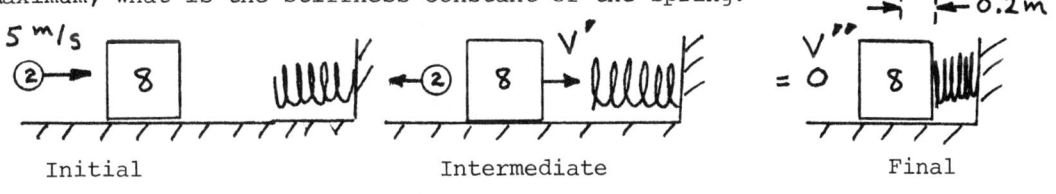

Initial   Intermediate   Final

***

Solve the collision problem by moving to the center of mass (CM) frame of reference.

$$V_{cm} = \frac{2 \times 5 + 8 \times 0}{2 + 8} = +1 \text{ m/s}$$

lab frame before: ②→ 5m/s   8   0 m/s

cm frame before: ②→ 4m/s   ←8 −1 m/s

cm frame after: ←② −2 m/s   8→ +½ m/s

lab frame after: ←② −1 m/s   8→ +1½ m/s

(In the above, velocity in the cm frame = velocity in lab − V_{cm}, and $u' = -\epsilon u$ in the cm frame.)

From the above the velocity of the 8 kg mass in the lab frame after the collision, V', is +1.5 m/s. All of the 8 kg block's kinetic energy goes into the potential energy of the spring, hence

$$\tfrac{1}{2} m V'^2 = \tfrac{1}{2} k s^2$$

$$k = \frac{m V'^2}{s^2} = \frac{(8 \text{ kg})(1.5 \text{ m/s})^2}{(0.2 \text{ m})^2}$$

$$\underline{\underline{k = 450 \text{ N/m}}}$$

6-33

A 20 g bullet moving initially at 600 (m/s), horizontally and to the right, becomes embedded in a wooden block of mass 4.5 kg moving initially in the opposite direction (to the left) at 8 m/s. Assuming that the collision is "head-on", so that the block-bullet system continues to move along the same horizontal line immediately after collision (whether to right or left), answer the following.

(a) Find the speed and direction of the block-bullet system immediately after collision.
(b) Can you apply momentum conservation to this problem? Briefly justify your response.
(c) Can you apply kinetic energy conservation to this problem?

****

BEFORE:
$v = 600$ m/s, $m = 20$ g (→)
$v = 8$ m/s, $M = 4.5$ kg (←)

AFTER:
$V' = ?$

(a) $P_i = P_f$ since $F_{ext} = 0$

$$mv + MV = (m+M)V'$$

$$V' = \frac{mv + MV}{m + M}$$

The block-bullet speed after collision

$$V' = \frac{0.02 \times 600 - 4.5 \times 8}{4.52} = -5.3 \text{ m/s}$$

(b) Yes $F_{ext} = 0$

(c) No.   Inelastic collision (In fact, the objects stick after collision; "perfectly inelastic")

$$\begin{cases} \text{Can see directly that } K_i \neq K_f \\ \frac{1}{2}mv^2 + \frac{1}{2}MV^2 \neq \frac{1}{2}(m+M)V'^2 \\ \text{From P conservation we saw} \\ V' = \frac{mv + MV}{m+M} \\ \text{Substitution on right side of the inequality proves the assertion} \end{cases}$$

6-34

A 0.05 kg bullet moving 300 m/s horizontally strikes a 2 kg target at rest on a smooth horizontal surface. The bullet passes all the way through the target, emerging with appreciable horizontal velocity. After the impact, the target moves horizontally (in the bullet's direction) with a velocity of 5 m/s.

(a) How fast is the bullet moving when it emerges from the target?
(b) Suppose it takes the bullet 2×10^{-4} sec to move through the target. What average force does it exert on the target?
(c) After this collision, what is the velocity of the center of mass of the system consisting of bullet and target?

* *

(a) The linear momentum P of the system doesn't change

$$P_{system} = \Sigma P_i = \Sigma m_i v_i$$
$$= mv(\text{Bullet}) + mv(\text{target})$$

$$P_{Before} = 0.05 \cdot 300 + 0 = 15 \text{ kg m/s}$$

$$P_{After} = 0.05 \cdot V + 2 \cdot 5$$

Equating: $0.05V + 10 = 15$
$V = 100$ m/s, speed of bullet after impact.

{ Linear momentum $\vec{P} = m\vec{v}$ is a vector, but here there is motion in only one dimension. }

{The kinetic energy of the system $K = K(bullet) + K(target)$ changes greatly because friction within the block dissipates it, producing thermal energy (heat)
Before: $K = \frac{1}{2} \cdot (0.05)(300)^2 = 2250$ J
After: $K = \frac{1}{2}(0.05)(100)^2 + \frac{1}{2}(2)(5)^2 = 275$ J
so it is INCORRECT to equate these.}

(b) Average force on bullet = $\dfrac{\text{change of bullet's momentum}}{\text{time to produce the change}}$

$\Delta p = mv_{After} - mv_{Before} = 0.05(100-300) = -10$ kg·m/s

$\bar{F} = -10/2 \times 10^{-4} = -5 \times 10^4$ N ($\approx$ 5 tons!)

The negative sign indicates force opposite to initial velocity.

{In this time, the bullet moved about $\frac{1}{2}(300+100)$ m/s $\times 2 \times 10^{-4}$ s = 4×10^{-2} m; the target was about 4 cm thick. Why is this just an estimate, not an exact value?}

(c) The system's momentum $P = MV_{cm}$ where M is the total mass ($0.05 + 2.0 = 2.05$ kg) and V_{cm} is the velocity of the center of mass. In (a) the system's momentum was found to be 15 kg·m/s (either before or after impact!), so
$2.05\, V_{cm} = 15 \quad \Rightarrow \quad V_{cm} = 7.32$ m/s

6-35

The coefficient of restitution of a certain "super ball" is 0.95. If the ball is dropped from a given height, how many bounces with the floor can it make before it fails to rebound to at least half of the original height from which it was dropped. (Assume the floor was the standard used when the coefficient of restitution was determined.)

****

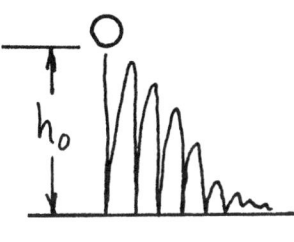

From the conservation of energy
$$mgh = \tfrac{1}{2}mv^2$$
So that the ball's speed just before it hits the floor the first time is
$$v_0 = \sqrt{2gh_0}$$

After the first collision with the floor its speed becomes $v_1 = \epsilon v_0$, where ϵ is the coefficient of restitution. Hence the height to which the ball rebounds is
$$h_1 = \frac{v_1^2}{2g} = \frac{(\epsilon v_0)^2}{2g} = \epsilon^2 h_0$$

Similarly after the second bounce $h_2 = \epsilon^2 h_1$.

In terms of the original height $h_2 = \epsilon^2(\epsilon^2 h_0)$.

Thus after n bounces $h_n = (\epsilon^2)^n h_0$

So we must calculate n such that $h_n = h_0/2$ or
$$h_n = \frac{h_0}{2} = \epsilon^{2n} h_0 \implies \tfrac{1}{2} = \epsilon^{2n}$$

Taking logs (either natural or decade) gives
$$\log(\tfrac{1}{2}) = 2n \log \epsilon$$
$$n = \tfrac{1}{2} \frac{\log(1/2)}{\log(\epsilon)} = \tfrac{1}{2} \frac{\log(.5)}{\log(.95)} = 6.76$$

After 7th bounce the rebound height will be less than $h_0/2$.

6-36

The figure below shows a glancing collision which is not necessarily elastic. Find the final velocity <u>vector</u> of the 6.0 kg mass.

Before: 1.0 kg mass with $v_1 = 10 \text{ m/s } \hat{\imath}$ approaching 6.0 kg mass at rest ($v_2 = 0$).

After: 1.0 kg mass moving at $|v_1'| = 5 \text{ m/s}$ at $36.87°$ above horizontal; 6.0 kg mass moving at angle $\alpha = ?$ below horizontal with $|v_2'| = ?$

***

Employ the conservation of (vector) momentum

$$p_{x \text{ initial}}^{\text{Total}} = m_1 v_{1x} + m_2 v_{2x} = (1)(10) + (6)(0) = 10$$

$$p_{y \text{ initial}}^{\text{Total}} = 0$$

$$p_{x \text{ final}}^{\text{total}'} = m_1 v_{1x}' + m_2 v_{2x}' = (1)(5)(\cos 36.87) + 6 v_2' \cos \alpha$$

$$p_{y \text{ final}}^{\text{Total}'} = m_1 v_{1y}' + m_2 v_{2y}' = (1)(5)(\sin 36.87) - 6 v_2' \sin \alpha$$

Equating $p_{x \text{ initial}}^{\text{Total}} = p_{x \text{ final}}^{\text{Total}}$ and $p_{y \text{ initial}}^{\text{Total}} = p_{y \text{ final}}^{\text{Total}}$

$$10 = 4 + 6 v_2' \cos \alpha \quad \text{or} \quad 1 = v_2' \cos \alpha$$
$$0 = 3 - 6 v_2' \sin \alpha \quad \quad\quad 1 = 2 v_2' \sin \alpha$$

Solving we get $\tan \alpha = 1/2$; $\underline{\underline{\alpha = 26.57°}}$

$$v_2' = \frac{1}{\cos \alpha} = \underline{\underline{1.12 \text{ m/s}}}$$

VARIABLE MASS SYSTEMS

6-37

Consider a box of mass 5kg. sliding across a frictionless surface with a speed of 10m/s. Rain is falling straight down into the box at the rate of 10 gm./s. . What is the speed of the box after one minute?

****

We will use cons. of momentum:

The initial momentum is $mv = 5 \cdot 10 = 50 \: \frac{kg \: m}{s}$

After one minute let its speed be v and therefore its momentum is

$$(5 + .6)v$$

where the .6 is the mass of the rain in the box after one minute

$$\therefore \quad 50 \: \frac{kg \: m}{s} = 5.6 \: v$$

$$8.9 \: \frac{m}{s} = v$$

6-38

Let us consider an unpowered cargo train, riding frictionlessly along a level track, which has its upper surface open to the elements of nature. Furthermore, let us create a scenario in which a storm is continuously increasing the mass of the train in the form of water at a rate of dm/dt, where m and t denote the rain's mass and time, respectively. If the initial mass and velocity of the train, before the rain started falling, is given by the symbols m_0 and v_0, respectively, find the position (equation of motion) of the train x(t), at a later time t assuming initially t=0 and x=0.

****

At $t=0$, the initial momentum of the train is

(1) $$p_0 = m_0 v_0$$

Let us denote the mass flux (rate of change of the mass with respect to time) by R, i.e.

(2) $$R \equiv \frac{dm}{dt} = \dot{m} = \text{constant}$$

such that the momentum of our train-rain system at a later time t can be written as

(3) $$p(t) = m_0 v(t) + R v(t) t = (m_0 + Rt) v(t)$$

From the conservation of momentum we may set expression (1) equivalent to that of (3), i.e. $p_0 = p(t)$ or

(4) $$m_0 v_0 = (m_0 + Rt) v(t)$$

Since the train's velocity may be written in the form:

(5) $$v(t) = \frac{d(x(t))}{dt} = \frac{dx}{dt}$$

then (4) becomes

(6) $$m_0 v_0 = (m_0 + Rt)(dx/dt)$$

Solving equation (6) for dx, gives us

$$dx = [(m_0 v_0)/(m_0 + Rt)] \cdot dt$$

and integration yields the desired result,

(7) $$x(t) = \int_0^x dx = m_0 v_0 \int_0^t \frac{dt}{(m_0 + Rt)} = \frac{m_0 v_0}{R} \ln\left(\frac{m + Rt}{m}\right)$$

$$= \frac{m_0 v_0}{R} \ln\left(1 + \frac{Rt}{m}\right)$$

7
ROTATIONAL DYNAMICS

ANGULAR DISPLACEMENT, VELOCITY, AND ACCELERATION

━━━━━━━━━━━━━━━━━━━━━━━━━━━━ 7-1

A motor is turning at 900 rpm. What angular acceleration will raise its angular velocity to 3000 rpm while it turns through 150 revolutions?

****

CONVERTING REVOLUTIONS TO RADIANS ONE HAS:

$$\omega_i = 900 \frac{rev}{min} \times \frac{1\ min}{60\ sec} \times \frac{2\pi\ rad}{rev} = 30\pi\ rad/sec$$

$$\omega_f = 3000 \times \frac{1}{60} \times 2\pi = 100\pi\ rad/sec.$$

$$\theta = 150\ rev \times 2\pi\ rad/rev = 300\pi\ rad.$$

SOLVE $\omega_f^2 = \omega_i^2 + 2\alpha\theta$ FOR α GIVING

$$\alpha = \frac{\omega_f^2 - \omega_i^2}{2\theta} = \frac{(100\pi)^2 - (30\pi)^2}{2 \times 300\pi} = 47.6 \frac{rad}{sec^2}$$

7-2

A flywheel of radius 30 cm starts from rest and accelerates with a constant angular acceleration of 0.50 rad/sec².

a) Compute the tangential acceleration and the radial acceleration of a point on the rim after it has turned through 270 degrees.
b) Compute the angular velocity at this same instant.
c) Compute the angular velocity after 6 seconds.
d) Compute the average angular velocity between the third and fourth second.

****

a) $a_T = \alpha r = (.50 \text{ rad/sec}^2)(30 \text{ cm}) = 15 \text{ cm/sec}^2$

$a_R = \dfrac{v^2}{r} = \dfrac{(r\omega)^2}{r} = r\omega^2 = r(\omega_0^2 + 2\alpha\theta)$

$= 2\alpha\theta r = 2(.50 \text{ rad/sec}^2)\left(\dfrac{3\pi}{2} \text{ rad}\right)(30 \text{ cm})$

$= 45\pi \text{ cm/sec}^2 \doteq 141 \text{ cm/sec}^2 \doteq 1.4 \text{ m/sec}^2$

b) $\omega^2 = \omega_0^2 + 2\alpha\theta = 0 + 2(.50)\left(\dfrac{3\pi}{2}\right) = \dfrac{3\pi}{2} \dfrac{\text{rad}^2}{\text{sec}^2}$

$\omega = \sqrt{\dfrac{3\pi}{2}} \text{ rad/sec}$

c) $\omega = \omega_0 + \alpha t = 0 + .50 \text{ rad/s} \times 6 \text{ sec} = 3.0 \text{ rad/sec}$

d) $\omega_{av} = \dfrac{\Delta\theta}{\Delta t} = \dfrac{\theta_4 - \theta_3}{4 - 3 \text{ sec}} = \dfrac{\frac{1}{2}\alpha(4)^2 - \frac{1}{2}\alpha(3)^2}{1 \text{ sec}} \text{ rad},$

having used the fact that ω_0 and θ_0 are both zero.

$\omega_{av} = \frac{1}{2}(.50)(16-9) = \dfrac{7}{4} = 1.75 \text{ rad/sec}$

Alternately, $\omega_{av} = \dfrac{\omega_3 + \omega_4}{2} = \dfrac{\alpha t_3 + \alpha t_4}{2}$

$\omega_{av} = \dfrac{.50(3+4)}{2} = 1.75 \text{ rad/sec}$

7-3

Consider a wheel of radius 30 inches allowed to spin on a central axle. The wheel is accelerated uniformly from rest to an angular speed of 50 rad/s during an interval of 10 s.
 (a) What is the angular acceleration α during this interval?
 (b) How many complete revolutions has the wheel made at the end of this interval?
 (c) If the wheel were rolling along the ground instead of spinning on its axle, what would its linear speed be at the end of the 10 s interval?
 (d) What is the minimum coefficient of static friction between the wheel and the ground required to keep the wheel rolling instead of sliding with this angular acceleration?

****

(a) $\alpha = \Delta\omega / \Delta t$ FOR CONSTANT ACCELERATION
$\Delta\omega = 50$ RADS/s, $\Delta t = 10$ s $\Rightarrow$ $\boxed{\alpha = 5.0 \text{ RADS/s}}$

(b) FOR UNIFORM ANGULAR ACCELERATION:
$$\Delta\theta = \omega_0 t + \tfrac{1}{2}\alpha t^2$$

$\omega_0 = 0$, $\alpha = 5$ RADS/s, $t = 10$ s
$\Rightarrow \Delta\theta = 250$ RADS

1 RAD = $1/2\pi$ REVS $\Rightarrow$ $\boxed{39 \text{ COMPLETE REVS.}}$

(c) IF WHEEL IS ROLLING:
$v = \omega r = 1500$ in/s
$\boxed{v = 125 \text{ ft/s}}$

(d) FOR ROLLING WHEEL, TRANSLATIONAL ACCELERATION GIVEN BY:
$$a = \alpha r = 12.5 \text{ ft/s}^2$$

THE FORCE RESPONSIBLE FOR THIS ACCELERATION IS THE STATIC FRICTION, f. SO:
$$f = ma$$
TO AVOID SLIPPING, MUST HAVE:
$$f < f_{MAX} = \mu_s mg$$
SO NEED: $\mu_s mg > ma$
$\Rightarrow \mu_s > a/g$ $\boxed{\mu_s^{MIN} = 0.39}$

7-4

When power is turned off a flywheel of radius 0.3 m, its angular speed decreases from 900 rev/min to 600 rev/min in 20 seconds. Assume uniform acceleration and leave factors of π indicated rather than multipled in:

(a) For how long will the flywheel continue to turn?
(b) In total, how many revolutions will the flywheel make while coming to rest after the power is turned off?
(c) Just after the power is turned off, what is the tangential acceleration, in m/s², of a point on the rim of the flywheel?
(d) Just after the power is turned off, what is the centripetal acceleration of a point on the rim?

* *

Describe the rotation by the angle θ between a "spoke" of the wheel and an arbitrary, fixed reference direction. The wheel's angular velocity is $\omega = d\theta/dt$ and its angular acceleration is $\alpha = d\omega/dt$. For constant α, integration gives $\omega = \omega_0 + \alpha t$ and $\theta = \theta_0 + \omega_0 t + \frac{1}{2}\alpha t^2$, where zero subscripts are values at $t = 0$. Elimination of t gives $\omega^2 = \omega_0^2 + 2\alpha(\theta - \theta_0)$ and elimination of α gives $\frac{1}{2}(\omega + \omega_0) = (\theta - \theta_0)/t$, which is the average angular velocity during time t. {These equations are analogous to those found for constant linear acceleration.}

(a) α describes how rapidly the velocity changes, so we first find it from $\omega = \omega_0 + \alpha t$: $\omega_0 = 900$ rev/min $= 15$ rev/s
at $t = 20$ s, $\omega = 600$ rev/min $= 10$ rev/s
$10 = 15 + 20\alpha \Rightarrow \alpha = -\frac{1}{4}$ rev/s²

Now find t for $\omega = 0$:
$0 = 15 + (-\frac{1}{4})t \Rightarrow t = 4 \times 15 = 60$ s

The wheel rotates for 60 s after turn-off or for 40 s ($= 60 - 20$) after its speed had dropped to 600 rev/min.

(b) $\theta - \theta_0 = \omega_0 t + \frac{1}{2}\alpha t^2 = 15\frac{\text{rev}}{\text{s}} \times 60 \text{s} + \frac{1}{2}(-\frac{1}{4}\frac{\text{rev}}{\text{s}^2})(60\text{s})^2 = 450$ rev
{OR: $\theta - \theta_0 = \frac{1}{2}(\omega + \omega_0)t = \frac{1}{2}(0 + 15) \cdot 60 = 450$ rev}

Tangential and centripetal accelerations are the components of the linear accelerations produced by the rotation. To calculate them we require <u>radian</u> measure for all angular quantities since only then can we write:

Arc length, $s = r\theta$ and linear speed, $v = ds/dt = r\,d\theta/dt = r\omega$, from which

$$\text{tangential acceleration } a_T = dv/dt = r\,d\omega/dt = r\alpha$$

and centripetal acceleration $a_c = v^2/r = r\omega^2$

∴ Convert: $\omega_0 = 15\,\text{rev/s} \times \dfrac{2\pi\,\text{rad}}{1\,\text{rev}} = 30\pi\,\text{rad/s}$, and

$\alpha = -\dfrac{1}{4}\,\text{rev/s}^2 \times \dfrac{2\pi\,\text{rad}}{1\,\text{rev}} = -\dfrac{\pi}{2}\,\text{rad/s}^2$

(c) At the rim, $r = 0.3\,\text{m}$, so

$$a_T = r\alpha = 0.3 \times (-\pi/2) = -0.15\pi\,\text{m/s}^2,$$

the negative sign showing that the linear speed is decreasing.

(d) At turn-off, $\omega = \omega_0$ and

$$a_c = r\omega_0^2 = 0.3(30\pi)^2 = 270\pi^2\,\text{m/s}^2,$$

directed from the point on the rim towards the axis.

ROTATION WITH CONSTANT ANGULAR FREQUENCY

7-5

A 500 gram mass is attached to a string and swung in a circle which is in the horizontal plane. The string makes an angle of 30° with the vertical. If the string is 1 m long, find the tension in the string and the angular velocity of the mass.

****

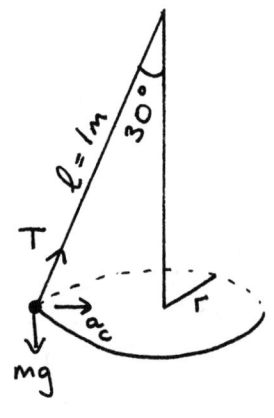

$$T\cos 30° = mg$$
$$\therefore T = \frac{mg}{\cos 30°} = \frac{(.5\,kg)(9.8\,m/s^2)}{\cos 30°} = 5.66\,N$$

$$T\sin 30° = ma_c = mr\omega^2$$
$$\therefore \omega = \sqrt{\frac{T\sin 30°}{mr}} = \sqrt{\frac{(5.66)N\,\sin 30°}{(.5\,kg)(1m\,\sin 30)}}$$

$r = \ell \sin 30 = 1\sin 30$
$\therefore r = .5\,m$

$\therefore \omega = 3.36$ rads/s

MOMENT OF INERTIA

7-6

A thin uniform rod has a linear mass density which is directly proportional to the distance from one end. (a) Find the mass of the rod. (b) Find the center of mass. (c) Find the moment of inertia about the axis shown. (d) Find the radius of gyration.

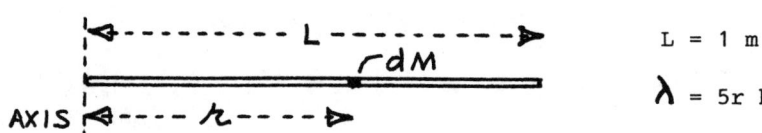

$L = 1$ m

$\lambda = 5r$ kg/m

****

(a) $M = \int dM = \int \lambda\,dr = \int_0^L 5r\,dr = \left.\frac{5r^2}{2}\right]_0^L = \frac{5L^2}{2}$

$M = 2.5$ Kg

(b) $X_{cm} = \int \frac{r \, dM}{M} = \int_0^L \frac{5r^2 \, dr}{2.5} = 0.667 \, m$

(c) $I = \int r^2 \, dM = \int_0^L 5r^3 \, dr = 1.25 \, Kg \, m^2$

(d) $MR^2 = I$, $R = \sqrt{\frac{I}{M}} = \sqrt{\frac{1.25}{2.5}} = 0.707 \, m$

--7-7

Calculate the moment of inertia of a rod about an axis through its end as shown. The rod's mass is m, its radius r and length L.

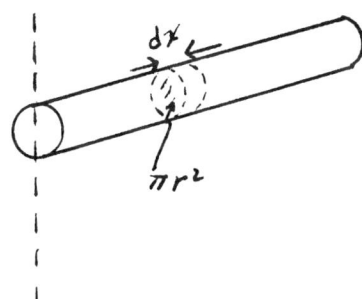

****

$I = \int x^2 \, dm$

But $dm = \rho \, dV$ (ρ the density)

$\qquad = \rho \pi r^2 \, dx$

$\therefore I = \int_0^L x^2 \rho \pi r^2 \, dx = \pi \rho r^2 \int_0^L x^2 \, dx = \pi \rho r^2 \left. \frac{x^3}{3} \right|_0^L$

$\qquad = \pi \rho r^2 \frac{L^3}{3}$

But $\rho = \frac{M}{V} = \frac{M}{\pi r^2 L}$ $\therefore I = \frac{ML^2}{3}$

7-8

The radius of the solid cylinder is 4 cm. Its mass is M grams.
(a) Find the moment of inertia of the cylinder about the YY' axis, when in position A.
(b) Find the distance x such that the moment of inertia about YY' is twice as great for the cylinder in position B as in position A.
The diagram is not drawn to scale.

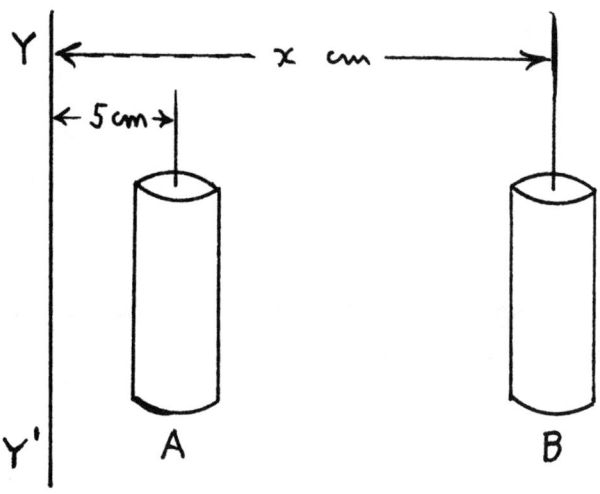

****

(a) Parallel Axis Theorem: $I = I_o + mb^2$

Hence $I_y = \frac{1}{2}mr^2 + mb^2$

$= \frac{1}{2}M(4)^2 + M(5)^2 = \underline{\underline{33M \text{ gm cm}^2}}$

(b) $2 I_y = I_o + mx^2$

$2(33M) = \frac{1}{2}M(4)^2 + Mx^2$

$66 = 8 + x^2 \qquad x^2 = 58 \qquad \underline{\underline{x = 7.6 \text{ cm}}}$

7-9

This symbol, ▰▰▰●m, shows a thin rod of mass M and length L, with a point mass particle, m, attached to one end. Two of these are connected so they rotate about the axis indicated by an X in the two diagrams below. In each case, find the moment of inertia, I, about the axis X.

a)

b)

****

I will use the fact that the moment of inertia of a thin rod about one end is $\frac{1}{3}ML^2$.

a) Adding the I's from the two m's and treating the combined rods as a single rod of length 2L,

$$I = mL^2 + m(2L)^2 + \frac{1}{3}(2M)(2L)^2$$

$$I = \left(5m + \frac{8}{3}M\right)L^2 \quad : \underline{\text{Answer}}$$

b) Adding up the four contributions,

$$I = mL^2 + mL^2 + \frac{1}{3}ML^2 + \frac{1}{3}ML^2$$

$$I = \left(2m + \frac{2}{3}M\right)L^2 \quad : \underline{\text{Answer}}$$

7-10

A nonuniform thin rod of length L lies along the x-axis with one end at the origin. It has a linear mass density λ kg/m given by

$$\lambda = \lambda_0 \sqrt{\frac{x}{L}}$$

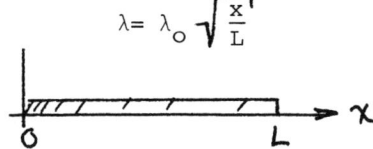

(a) Find the total mass.
(b) Find the center of mass.
(c) Find the moment of inertia if rotating about a perpendicular axis through $x = 0$.

****

(a) $M = \int_0^L dm = \int_0^L \lambda\, dx = \int_0^L \lambda_0 \sqrt{\frac{x}{L}} = \frac{\lambda_0}{\sqrt{L}} \int_0^L x^{1/2} dx$

$= \frac{\lambda_0}{\sqrt{L}} \cdot \frac{x^{3/2}}{3/2} \Big|_0^L = \underline{\underline{\tfrac{2}{3} \lambda_0 L}}$

(b) First moment $= \int_0^L x\, dm = \frac{\lambda_0}{\sqrt{L}} \int_0^L x^{3/2} dx = \tfrac{2}{5} \lambda_0 L^2$

$X_{cm} = \frac{\text{First moment}}{M} = \frac{2/5\, \lambda_0 L^2}{2/3\, \lambda_0 L} = \underline{\underline{\tfrac{3}{5} L}}$

(c) Moment of inertia $=$ second moment $= \int_0^L x^2 dm$

$= \frac{\lambda_0}{\sqrt{L}} \int_0^L x^{5/2} dx = \tfrac{2}{7} \lambda_0 L^3 = \frac{M \cdot 2/7\, \lambda_0 L^3}{2/3\, \lambda_0 L}$

$= \underline{\underline{\tfrac{3}{7} M L^2}}$

TORQUE AND ANGULAR ACCELERATION

7-11

An 8.00-kg block, resting on a frictionless surface, is attached to a light cord that is wrapped around a flywheel of radius 0.130 m.
The acceleration of the block down the incline is observed to be 2.40 m/s².
Find (a) the tension in the cord,
(b) the angular acceleration of the flywheel, and
(c) the flywheel's rotational inertia.

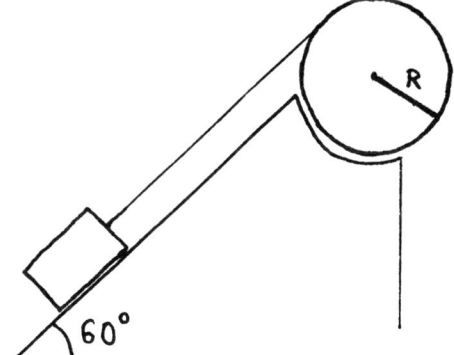

***

Since the problem seems to be about accelerations, forces, and inertia, the most natural way to attack it is using *dynamics* - rather than by using conservation of energy, which might be more suitable if velocities and distances were involved.

(a) "$F = ma$" for block:

$$mg \sin 60° - T = ma$$
$$T = m(g \sin 60° - a)$$

$$T = (8\,kg)\left[\left(9.81\,\tfrac{m}{sec^2}\right)\sin 60° - 2.40\,\tfrac{m}{sec^2}\right]$$

$$T = \underline{\underline{48.7\,N}}$$

(c) "$\tau = I\alpha$" for flywheel:

$$I = \frac{\tau}{\alpha} = \frac{TR}{\left(\frac{a}{R}\right)} = \frac{TR^2}{a}$$

$$I = \frac{(48.7\,kg\cdot m)(.130\,m)^2/sec^2}{(2.40\,m)/sec^2}$$

$$I = \underline{\underline{0.343\,kg\cdot m^2}}$$

(b) $\alpha = \dfrac{a}{R} = \left(\dfrac{2.4\,m}{sec^2}\right)\left(\dfrac{rad}{.13\,m}\right)$

$$\alpha = \underline{\underline{18.5\,rad/sec^2}}$$

7-12

It has been proposed that a space station be constructed in the form of an annular ring which may be set into rotation about its axis. The resulting centripetal acceleration produces an "artificial gravity", avoiding the problems of weightlessness. Such a space station has a mass of 40,000 kg and a radius of 20 m.

a) What must be the space station's angular velocity in order to have an "artificial gravity" of 9.8 m/s^2?

b) The space station is constructed in space. The required angular velocity is achieved by firing two rockets for 50.0 s. What must be the constant thrust of each rocket?

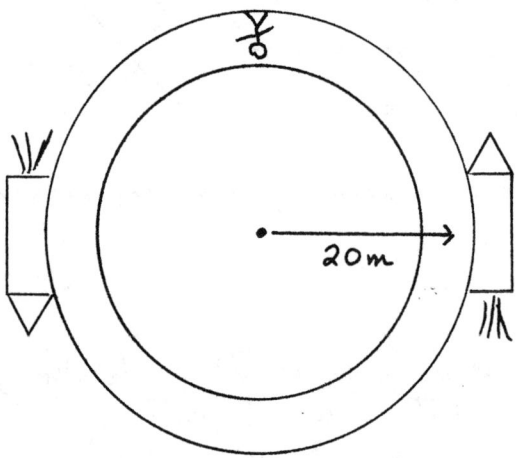

***

a) $a_c = \omega^2 r = g$; $\omega = \sqrt{g/r} = \sqrt{\dfrac{9.8 \text{ m/s}^2}{20 \text{ m}}} = \underline{0.70 \text{ rad/s}}$

b) Torque = τ $\qquad \tau = I\alpha \qquad I = Mr^2 \qquad \alpha = \dfrac{\Delta \omega}{\Delta t}$

$\tau = 2 F_T \cdot r \qquad (F_T = \text{thrust of each rocket})$

$\therefore \quad 2 F_T \cdot r = (Mr^2) \dfrac{\Delta \omega}{\Delta t}$

$F_T = \dfrac{M r \omega_f}{2 \Delta t} = \dfrac{4 \times 10^4 \text{ kg} \times 20 \text{ m} \times 0.7/\text{s}}{2 \times 50 \text{ s}} = \underline{5600 \text{ N}}$

7-13

Four point masses are placed on the surface of a disk, as shown in the figure. The 5-kg disk has a radius of 2.5 meters and is free to rotate about an axis passing through its center (perpendicular to the plane of the paper), but is initially held in place. Find the moment of inertia of this system, and its initial angular acceleration just after the disk is released.

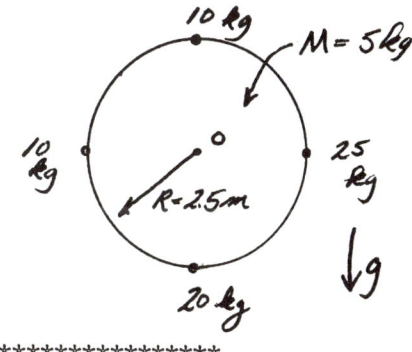

****

The moment of inertia of the entire system about the central axis O is the sum of the contributions from the disk ($\frac{1}{2}MR^2$) and from the point masses ($m_i R^2$)

$$\Rightarrow I_0 = \frac{1}{2} \cdot 5 kg \cdot (2.5 m)^2 + 10 kg \cdot (2.5m)^2 + 25 kg (2.5m)^2$$
$$+ 20 kg \cdot (2.5m)^2 + 10 kg (2.5m)^2 \Rightarrow \boxed{I_0 = 422\ kg\cdot m^2}$$

There are four forces on the disk (not counting its own weight) — one from each of the four point masses. The forces from the weights at the top and bottom of the disk exert no torque about O (their line of action passes through O $\Rightarrow$ zero lever arm).

The net torque comes from the masses on the left and right sides:

$$\Sigma \tau = 25 kg \cdot 9.8 m/s^2 \cdot 2.5 m - 10 kg \cdot 9.8 m/s^2 \cdot 2.5 m$$

$\ominus$ since in opposite direction

$$\Rightarrow \tau_{net} = 368\ N\cdot m = I_0 \cdot \alpha$$

$$\Rightarrow \boxed{\alpha = \tau_{net}/I = 0.871\ rad/sec^2}$$

7-14

The magnitude of the angular momentum of a rotating rigid body about a fixed axis is changing with time according to the formula $L = (3 - 4t + 5t^2)$ kg-m^2/sec. The time t is expressed in seconds.

a) What is the instantaneous torque on the body as a function of time?
b) What is the instantaneous torque on the body at $t = 0$ and $t = 2.0$ seconds?
c) What is the average torque during the interval 0 to 3.0 seconds?
d) If the moment of inertia of the body is a constant 10.0 kg-m^2, what is the angular velocity at the end of 4.0 seconds?
e) What is the angular acceleration at the end of 5.0 seconds, again assuming the constant moment of inertia in d) above?

***

a) $\tau = \dfrac{dL}{dt} = \dfrac{d}{dt}(3 - 4t + 5t^2) = -4 + 10t \; \dfrac{kg\,m^2}{sec^2}$ or Nt-m

b) $\tau_0 = -4 + 10(0) = -4$ Nt-m
$\tau_2 = -4 + 10(2) = 16$ Nt-m

c) $\tau_{av} = \dfrac{\Delta L}{\Delta t} = \dfrac{L_3 - L_0}{3 - 0\,sec} = \dfrac{[3 - 4(3) + 5(3)^2] - [3 - 4(0) - 5(0)^2]}{3 \, sec}$

$= \dfrac{(3 - 12 + 45) - 3}{3} = \dfrac{33}{3} = 11$ Nt-m

d) $L = I\omega$ so $\omega = \dfrac{L}{I} = \dfrac{3 - 4(4) + 5(4)^2 \; kg\,m^2/s}{10 \; kg\,m^2}$

$\omega = \dfrac{67}{10}$ rad/sec $= 6.7$ rad/sec

e) $\tau = I\alpha$ so $\alpha = \dfrac{\tau}{I} = \dfrac{-4 + 10(5) \; kg\,m^2/sec^2}{10 \; kg\,m^2}$

$\alpha = \dfrac{46}{10}$ rad/sec^2 $= 4.6$ rad/sec^2

ROTATION WITH CONSTANT ANGULAR ACCELERATION

7-15

A SPACE STATION IS CONSTRUCTED AS A LARGE CYLINDRICAL DISK OF RADIUS = 100 METERS AND HEIGHT = 10 METERS. ITS MASS IS 1.0E5 KILOGRAMS (ASSUME IT IS HOMOGENEOUS). TO INDUCE ARTIFICIAL GRAVITY, IT IS SET INTO ROTATION BY 2 JETS, EACH EXHAUSTING ITS GAS AT A 10 KILONEWTON FORCE TANGENTIAL TO THE EDGE.

HOW LONG DOES THE JET HAVE TO BURN FOR THE STATION TO REACH AN ANGULAR VELOCITY OF 1 RAD/SEC?

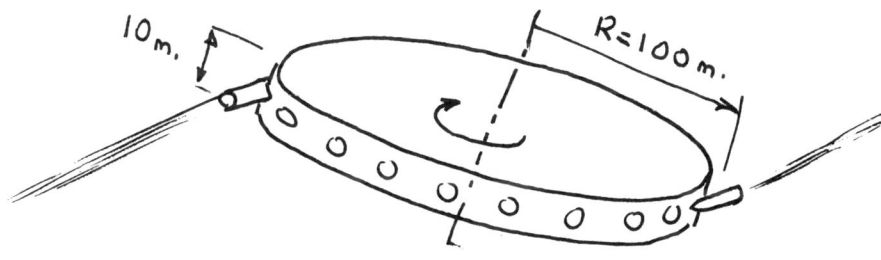

****

Time and angular velocity are related through the angular acceleration

$$\omega = \alpha t$$

Therefore, we must find the angular acceleration. It is in turn related to the applied force through the torque and moment of inertia

$$\alpha = \tau / I$$

But torque equals (both jets)

$$\tau = 2RF\sin\theta = 2 \times 100m \times 10kN \times 1 = 2E6\,Nm$$

where F is the force of the jet and R is the perpendicular distance of the jet force to the space station axis.

Moment of inertia for a solid cylindrical disk equals

$$I = mR^2/2 = 1.0E5 \text{ kg} \times (100m)^2/2 = \frac{E9}{2} \text{ kgm}^2$$

Therefore angular acceleration is found

$$\alpha = \frac{2E6 \text{ Nm}}{E9/2 \text{ kgm}^2} \times \frac{\text{kg m/s}^2}{N} = 4E-3 /\text{sec}^2$$

Solving for time

$$t = \omega/\alpha = \frac{1/\text{sec}}{4E-3/\text{sec}^2} = 2.5E2 \text{ sec.}$$

7-16

A disc of radius .2 m is mounted on a shaft of radius .1 m. This combination is mounted in a frictionless pivot and acted on by constant forces of 100 N and 50 N as shown below. The moment of inertia of the combination about the pivot is 3 kg-m^2. (a) What is the angular acceleration of the disc-shaft combination? (b) If the combination was released from rest, what is the angular velocity after 4 seconds? (c) How many revolutions does the disc-shaft combination make in 4 seconds? (d) What is velocity of point A and point B after 4 seconds?

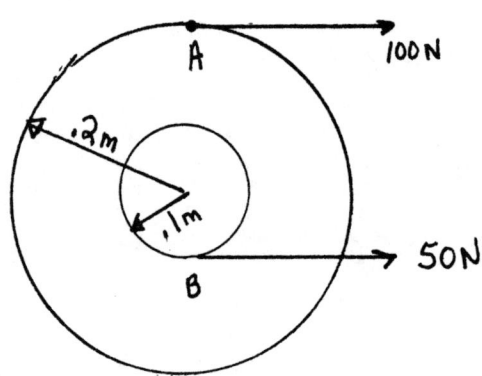

****

a) $\Sigma \tau = I\alpha$

$(100N)(.2m) - (50N)(.1m) = (3 kg\text{-}m^2)\alpha$

$\alpha = \dfrac{(20 N\text{-}m) - (5 N\text{-}m)}{3 kg\text{-}m^2} = 5 \, rad/s^2$

b) $\omega = \omega_0 + \alpha t = 0 + (5 \, rad/s^2)(4s) = 20 \, rad/s$

c) $\theta = \theta_0 + \omega_0 t + \tfrac{1}{2}\alpha t^2$

$\theta = 0 + 0 + \tfrac{1}{2}(5 \, rad/s^2)(4s)^2$

$= 40 \, rad \times \dfrac{1 \, rev}{2\pi \, rad} = 6.37 \, rev$

d) $v_{\parallel} = r\omega = (20 \, rad/s)(.2m) = 4 \, m/s$ at A.

$v_{\parallel} = (20 \, rad/s)(.1m) = 2 \, m/s$ at B.

7-17

A bicycle wheel of mass M = 2.5 kg and radius R = 0.70 m has a constant angular acceleration α = 5.0 (rad.)/s².

(a) Find the angular speed ω of the wheel 3.0 seconds after it has started from rest.

(b) Find the angular displacement $\Delta\theta$ of the wheel in that time, i.e., the total angle a spot on the wheel has turned through in the 3.0 second time interval.

(c) Assuming that the bicycle wheel may be approximated as a hoop of the same radius, what is the moment of inertia of the wheel about axis OO' through the center of the wheel and oriented perpendicularly to the plane of the wheel.

(d) Find the moment of inertia about an axis MM', parallel to axis OO' but shifted from the center of the hoop by a distance R/3. Distance OM in the figure is 1/3 R.

(e) Find the torque, referred to axis OO', required to produce the given angular acceleration, α = 5.0 (rad.)/s².

(f) Find the angular momentum of the hoop after the three seconds have passed. (Again take the axis to be OO'.)

****

(a) $\omega = \omega_0 + \alpha t = 0 + 5.0 \frac{(rad.)}{s^2} \times 3.0\,s = 15 \frac{(rad.)}{s}$

(b) $\theta = \theta_0 + \omega_0 t + \frac{1}{2}\alpha t^2 = 0 + 0 + \frac{1}{2} \times 5.0 \times (3)^2 = 23\,(radians)$

(c) $I_{OO'} = MR^2$ (Obvious from definition: $I = \sum m_i r_i^2$)

(d) $I_{MM'} = I_{OO'} + M(\frac{1}{3}R)^2 = MR^2 + \frac{1}{9}MR^2 = 1.1\,MR^2$

   Numerical Answers to (c) and (d):
   (c) $I_{OO'} = MR^2 = 1.2\,kg \cdot m^2$
   (d) $I_{MM'} = 1.1\,MR^2 = 1.4\,kg \cdot m^2$

(e) $\tau_{OO'} = I_{OO'}\alpha = MR^2\alpha = (2.5)(0.7)^2(5) = 6.1\,N \cdot m$

(f) $\ell = I\omega = MR^2 \omega$
   $\omega = 15 \frac{(rad.)}{s}$, [part (a)]

   $\therefore \ell = 18.5\,kg \cdot m^2/s$

7-18

A mechanical clock with a second hand that is 10 cm long keeps time accurately only when it is fully wound. (a) Just after it is wound, what is the centripetal acceleration of the tip of the second hand? (b) If it takes 24 hours for the second hand to come to a complete stop, what is the angular acceleration (assumed constant) of the second hand? (c) How many revolutions will the second hand have made two hours after being wound?

****

(a) $a_c = \dfrac{v^2}{R}$ because it will take 60 sec to make one complete revolution

$$v = \dfrac{2\pi R}{60} = \dfrac{2\pi (0.1)}{60} = 0.0105 \text{ m/sec}$$

$$a_c = \dfrac{(0.0105)^2}{0.1} = \underline{0.0011 \text{ m/sec}^2}$$

(b) In terms of the initial and final angular speeds (ω)

$$\omega_f = \omega_0 + \alpha t$$

$$\omega_0 = \dfrac{v}{R} = \dfrac{0.0105}{0.1} = .105 \text{ rad/sec}$$

$$\omega_f = 0 \Rightarrow \alpha = \dfrac{-\omega_0}{t} = \dfrac{-.105}{24(3600)}$$

$$= \underline{-1.22 \times 10^{-6} \text{ rad/sec}^2}$$

(c) Don't think that this is easy... remember that the second hand is slowing down. If we define the angular distance θ in radians:

$$\theta = \tfrac{1}{2}\alpha t^2 + \omega_0 t$$
$$= \tfrac{1}{2}(-1.22 \times 10^{-6})(2 \times 3600)^2 + (.105) \times (2 \times 3600)$$
$$= 724.4 \text{ rad}$$

Since there are 2π radians in each revolution:

$$\#Rev = \dfrac{\theta}{2\pi} = \underline{115.3 \text{ rev}}$$

7-19

A massless string is wrapped about a uniform, solid cylinder which rotates without friction about its fixed cylinder axis. The other end of the string is attached to a block which slides without friction on a plane surface inclined by angle θ to the horizontal. (See accompanying diagram.)

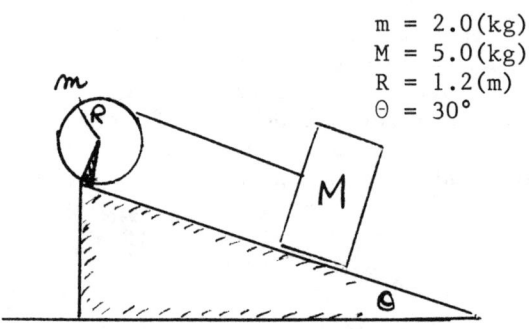

m = 2.0 (kg)
M = 5.0 (kg)
R = 1.2 (m)
θ = 30°

(a) Draw free-body diagrams for each of the two objects, the cylinder of mass m and the block of mass M.
(b) Using these as a guide, find the linear acceleration of mass M, the angular acceleration of the cylinder, and the tension in the string.

****

(a)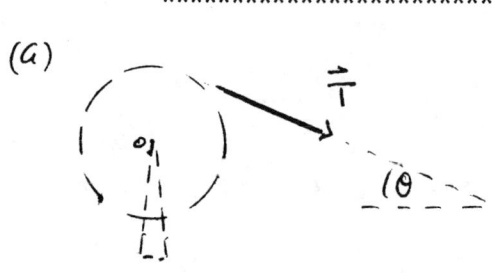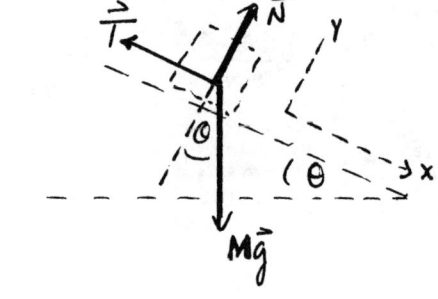

It is unnecessary to include forces which act at the axis of rotation

From the figure at left above, $\tau = I\alpha$
(Here τ is the torque, I the moment of inertia referred to point o, and α the angular acceleration.)

$\therefore TR = (\tfrac{1}{2} m R^2) \alpha$; $I = \tfrac{1}{2} MR^2$ for uniform cylinder about cylinder axis.

Eq(1) $\therefore \underline{T = \tfrac{1}{2} m a}$, where a is linear acceleration $(\alpha = \tfrac{a}{R})$.

From the figure above right, we have the y-component equation

Eq(2) $\underline{N - Mg \cos\theta = 0}$

and the x-component equation $Mg \sin\theta - T = Ma$, or

Eq(3) $\underline{T = M(g \sin\theta - a)}$

Equations (1) and (3) represent a solution for the two unknowns T and a. Eliminating T gives

$$\tfrac{1}{2} m a = M(g \sin\theta - a), \text{ or}$$

Eq 4
$$a = \frac{Mg \sin\theta}{M + \tfrac{1}{2} m}$$

Note that $a < g$

$$\therefore a = 4.1 \text{ m/s}^2$$

Since $a = R\alpha$, $\quad \alpha = \dfrac{a}{R} = \dfrac{4.1}{1.2} = \underline{3.4 \text{ s}^{-2}}$

To find T use Eq. 4 in either of the equations (1) or (3). Taking (1):

$$T = \tfrac{1}{2} m a = \tfrac{1}{2} m \left(\frac{M \sin\theta}{M + \tfrac{1}{2} m} g \right) = \tfrac{1}{2}(2.0)(4.1) = \underline{4.1 \text{ N}}$$

Note that $T = \left(\dfrac{m \sin\theta}{m + 2M}\right) Mg$. Owing to the rotational inertia of the cylinder, even if the incline were to be placed in a vertical position ($\theta = \pi/2$), $T < Mg$, the weight of the suspended mass.

ROTATION WITH TRANSLATION

7-20

A sphere rolls along a horizontal smooth surface without slipping at a constant linear speed. Show that its rotational kinetic energy about the center of mass is 2/7 of its total kinetic energy.

****

"Without slipping" means that $v_{CM} = R\omega$, where v_{CM} is the constant speed of the center of mass, R is the radius and ω is the angular speed, which is constant for an obvious reason.

Since the moment of inertia of a sphere about its center is $\frac{2}{5}MR^2$,

$$\frac{(K.E.)_{rotational}}{(K.E.)_{total}} = \frac{(K.E.)_{rot.}}{(K.E.)_{transl.} + (K.E.)_{rot.}}$$

$$= \frac{\frac{1}{2}\left(\frac{2}{5}MR^2\right)\omega^2}{\frac{1}{2}Mv_{CM}^2 + \frac{1}{2}\left(\frac{2}{5}MR^2\right)\omega^2} = \frac{\frac{2}{5}v_{CM}^2}{\left(1+\frac{2}{5}\right)v_{CM}^2} = \frac{2}{7}$$

$$\therefore (K.E.)_{rot} = \frac{2}{7}(K.E.)_{total} \leftarrow$$

7-21

A circular hoop of mass m and radius R is originally spinning about a horizontal axis through its center at a rotation rate of 10.0 radians/sec. The hoop is then placed on the floor. At first the hoop spins in one place, then, due to friction, it begins to move forward while still slipping. Finally the hoop begins to roll without slipping. Calculate the hoop's angular speed when it begins to roll without slipping.

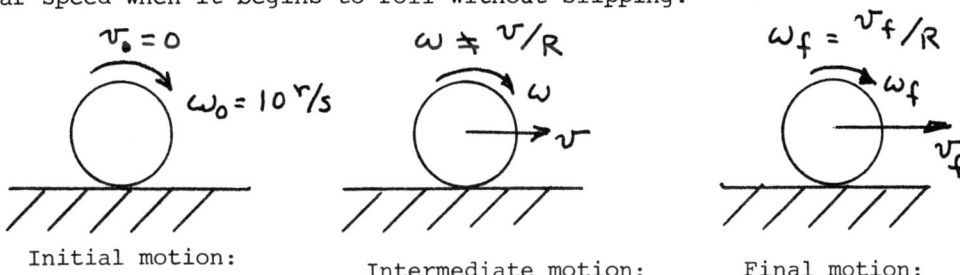

Initial motion: no translation

Intermediate motion: translation and rotation with slipping

Final motion: no slipping

****

Use the convention that motion to the right is positive and that the commensurate clockwise rotation is positive. Separate the effect of the frictional force on the hoop into (1) changes in the motion of the center of mass (CM) and (2) changes in rotation about the CM. Note that the force of friction on the hoop edge where it touches the floor is to the right, hence positive. However, the torque due to the frictional force about the CM is anti-clockwise, hence **negative**. Use the impulse formulation for Newton's 2nd law:

(1) $F_{fric} \Delta t = m v_{final} - m v_{initial}$ (linear)

(2) $- R F_{fric} \Delta t = I \omega_{final} - I \omega_{initial}$ (rotat.)

Divide equation (2) by (1) recalling $v_{initial} = 0$ and $I = mR^2$ for a hoop.

$$-R = \frac{I\omega_f - I\omega_0}{m v_f} = \frac{R^2(\omega_f - \omega_0)}{v_f}$$

The "no slip" condition is $v_f / R = \omega_f$, hence

$$-R^2 \omega_f = R^2 \omega_f - R^2 \omega_0$$

or $\quad \omega_f = \omega_0 / 2 = 5$ radians/second.

ANGULAR MOMENTUM AND IMPULSE

7-22

A 5-kg rod is 1.5 meters long and pivoted about its center. The rod is struck at one end with a hammer; during the blow, an average force of 1000 newtons acts for 0.01 seconds. What is the angular velocity of the rod just after impact, if the rod was initially at rest?

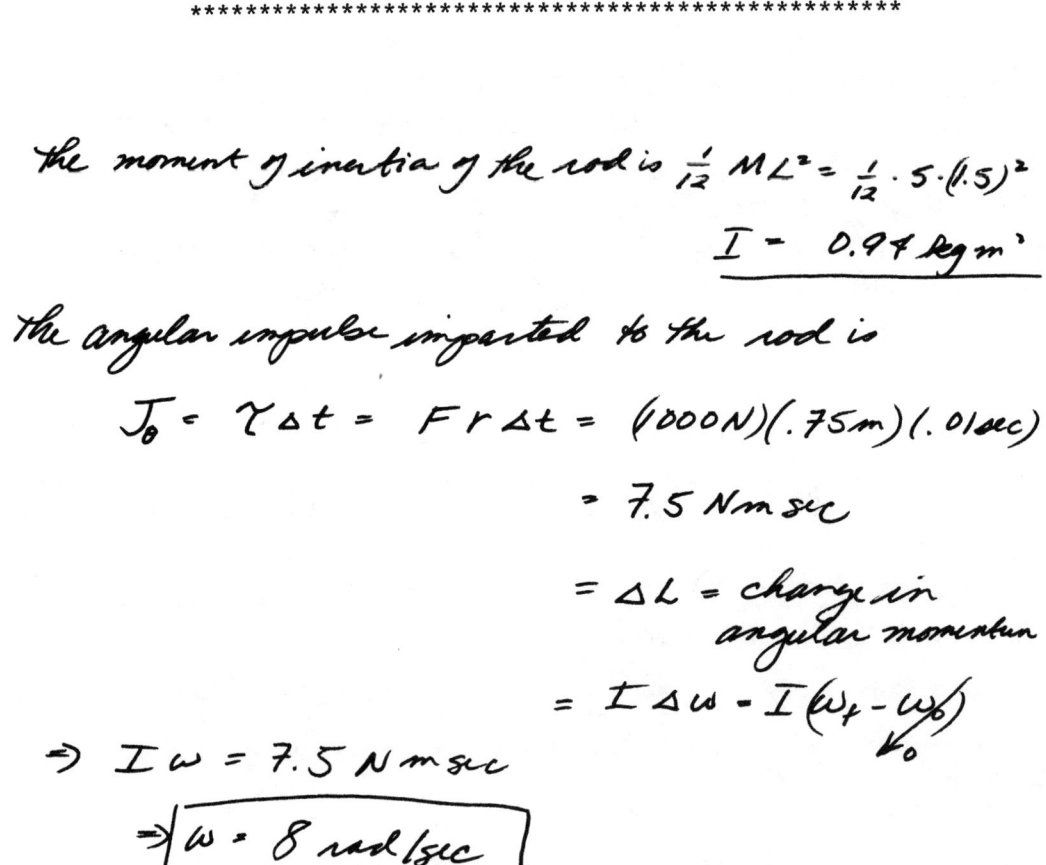

***

The moment of inertia of the rod is $\frac{1}{12}ML^2 = \frac{1}{12} \cdot 5 \cdot (1.5)^2$

$$I = 0.94 \text{ kg m}^2$$

The angular impulse imparted to the rod is

$$J_\theta = \tau \Delta t = Fr\Delta t = (1000N)(.75m)(.01 sec)$$

$$= 7.5 \text{ Nm sec}$$

$$= \Delta L = \text{change in angular momentum}$$

$$= I\Delta\omega = I(\omega_f - \omega_0)$$

$$\Rightarrow I\omega = 7.5 \text{ Nm sec}$$

$$\boxed{\Rightarrow \omega = 8 \text{ rad/sec}}$$

7-23

A phonograph turntable has a moment of inertia of 0.04 kg-m^2. The table is set to run at 45 rpm. A 2 gram coin is placed on the turntable a distance from the axis of rotation. The coefficient of friction (static and kinetic) between coin and table is 0.3.

a) Find the maximum distance the coin may be placed from the axis without slipping radially.
b) The coin, when dropped at this radius is observed to slip tangentially for a short distance and time. Find the time interval of the tangential slip.

****

a) $\dfrac{mv^2}{R} = mR\omega^2 = \mu mg$

$\omega = 45 \dfrac{\text{rev}}{\text{min}} \times 2\pi \dfrac{\text{rad}}{\text{rev}} \times \dfrac{1}{60} \dfrac{\text{min}}{\text{s}} = 4.71 \text{ s}^{-1}$

$R = \dfrac{\mu g}{\omega^2} = \dfrac{0.3 \times 9.8}{(4.71)^2} = 0.133 \text{ m}$
$= 5.2 \text{ inch}$

b) $\tau \Delta t = \Delta L$ or $F \Delta t = \Delta p$

$\mu mg R \Delta t = mR^2 \omega$ $\mu mg \Delta t = mv = mR\omega$

$\Delta t = \dfrac{R\omega}{\mu g} = \dfrac{0.133 \times 4.71}{0.3 \times 9.8} = 0.21 \text{ s}$

CONSERVATION OF ANGULAR MOMENTUM

7-24

A carrousel of radius 4.8 m is at rest but free to rotate. Its moment of inertia is 75kg-m^2. A 40kg youngster running at 5.0 m/s in a direction tangent to the rim of the carrousel jumps on and stops (with respect to the carrousel). Find the resulting angular velocity of the system.

****

The child and carrousel form an isolated system with no external torques acting, so conservation of angular momentum holds.

Angular momentum before = mvr = 40 × 5 × 4.8

= 960 kg-m^2

Angular momentum after = $(I_c + mR^2)\omega$

= (75 + 40 × 4.8^2)ω

So

960 = 267 ω

ω = 3.6 rad/s

7-25

A man with mass 80 kg runs around the edge of a horizontal turntable mounted on a vertical frictionless axis through its center. The velocity of the man relative to the earth is 2.0 m/s. The turntable is rotating in the opposite direction with an angular velocity of 0.2 rad/sec. The radius of the turntable is 3.0 meters and its moment of inertia about the axis of rotation is 800 kg-m². Calculate the angular velocity of the system if the man comes to rest relative to the turntable.

***

Angular momentum will be conserved because there are no external torques acting along the axis of rotation. However, kinetic energy is $\underline{not}$ conserved.

$$L_{TOT} = I_{man} \omega_{man} + I_{TT} \omega_{TT}$$

$$= Mr^2 \left(\frac{v}{r}\right) + 800(-0.2) \; kg\,m^2/s$$

$$= (80\,kg)(2.0\,m/s)(3.0\,m) - 160 \; kg\,m^2/s$$

$$= 480 - 160 = 320 \; kg\,m^2/s$$

$$L_{TOT} = L_{final} = (I_{man} + I_{TT}) \omega_f = 320 \; kg\,m^2/s$$

$$= (Mr^2 + 800) \omega_f = [80(3.0)^2 + 800] \omega_f$$

$$= (720 + 800) \omega_f = 1520 \, \omega_f \; kg\,m^2/s$$

$$\omega_f = \frac{L_{TOT}}{1520 \; kg\,m^2} = \frac{320}{1520} \; rad/s = .21 \; rad/sec$$

in the same direction as the man was originally running.

7-26

An apparatus to determine the speed of a rifle bullet consists of a ½ kg can mounted on the edge of a horizontally mounted disk which has a mass of 5 kg and a radius of 11 cm. The bullet, which has a mass of 1.5 g, lodges in the center of the can a distance 16 cm from the axis of rotation of the disk. Initially the disk with can is at rest. After the impact of the bullet, the disk with can and bullet rotates with an angular velocity of 2.03 rad/s. What was the initial velocity of the bullet before it lodged in the can? (Hint: approximate can as a point mass located at its center of mass.)

****

Describing problem:

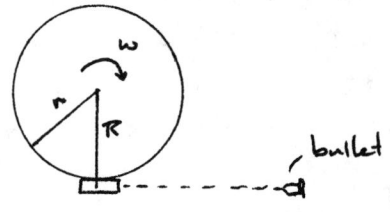

m_c (can) = 0.5 kg
M (disk) = 5 kg
m_b (bullet) = 1.5 g
R = 16 cm
r = 11 cm
ω = 2.03 rad/s

Setting up eqs.

$L_i = I_b \omega_b = (m_b R^2)(v/R) = m_b v R$

$L_f = I_D \omega + I_c \omega + I_b \omega = \left(\tfrac{1}{2} M r^2 + m_c R^2 + m_b R^2\right) \omega$

Equating $L_i = L_f$

$\Rightarrow m_b v R = \left(\tfrac{1}{2} M r^2 + m_c R^2 + m_b R^2\right) \omega$

$\Rightarrow v = \dfrac{\left(\tfrac{1}{2} M r^2 + m_c R^2 + m_b R^2\right) \omega}{m_b R}$

Substituting numbers:

$v = \dfrac{\left[\tfrac{1}{2}(5)(.11)^2 + (.5)(.16)^2 + (.0015)(.16)^2\right]}{(.0015)(.16)} (2.03)$

$= \underline{\underline{364 \text{ m/s}}}$

7-27

A 2 kg rod of length 100 cm is suspended freely from one end. A 0.5 kg mass moving at 40 m/s hits the rod at its center of mass and sticks there. Compute the angular speed of the system just after the collision.

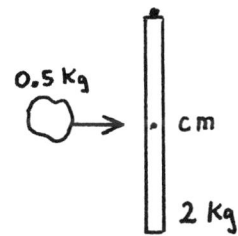

****

I (rod) about one end
$$= \frac{ML^2}{3}$$
$L = 1\ m$
$r = .5\ m$

before   after

the angular momentum about the axis is conserved

$$L_{before} = L_{after}$$

$$mvr = I_{total}\ \omega = [I_{mass} + I_{rod}]\ \omega$$

$$(.5)(40)(.5) = \left[.5(.5)^2 + 2\frac{(1)^2}{3}\right]\omega$$

$$\omega = 12.6\ rad/s$$

264 / Rotational Dynamics

7-28

A cylindrical satellite of radius 1.5 m, moment of inertia 800 kg m^2, spins at the rate of 1 revolution every 6 seconds. If an astronaut grabs a handhold at the side of the satellite, and if her net mass of 100 kg is centered about 2 meters from the spin axis, what will be the new spin rate, approximately?

****

Initial angular momentum: $L_1 = I\omega_1 = 800 \text{ kg m}^2 \times \dfrac{1 \text{ rev}}{6 \text{ sec}} = \dfrac{800}{6}$ (weird units!)

Final angular momentum:

$L_2 = (I + mr^2)\omega_2 = (800 + 100 \times 2^2) \text{ kg m}^2 \times \omega_2 = 1200\, \omega_2$

By conservation of angular momentum, $L_1 = L_2$,

$1200\, \omega_2 = \dfrac{800}{6}$, $\quad \underline{\omega_2 = \dfrac{1 \text{ rev}}{9 \text{ sec}}}$

KINETIC ENERGY, WORK AND POWER

7-29

A mass of 4 kg attached to a rope wound about a drum of radius 0.15 m and moment of inertia 0.18 kgm^2 about its axis of rotation. There is no friction. The 4 kg mass is released from rest.
a) Calculate the acceleration of the 4 kg mass.
b) Calculate the tension in the rope.
c) Calculate the kinetic energy of the drum when the 4 kg mass has fallen 2.5 meters.

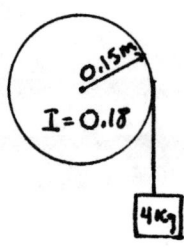

****

a) Draw Two Free Body Diagrams:

Drum — (circle, radius .15, with α down arrow, T down arrow)

"Mass" — 4kg block, T' up, a down, $4(9.8) = 39.2 N$ down

assume massless rope $\Rightarrow T = T'$

also $\alpha = \dfrac{a}{r} = \dfrac{a}{0.15}$

assume clockwise positive and down positive

Torque Eq. For Drum: $T(0.15) = 0.18\alpha = \dfrac{0.18a}{0.15}$

$\Sigma F_{vert} = ma$ for mass: $39.2 - T = 4a$

$\Rightarrow T = \dfrac{0.18a}{(0.15)^2} = 8a \qquad \Rightarrow 39.2 - 8a = 4a$

$$a = 3.27 \, m/s^2$$

b) $T = 8a = 26.1 N$

c) use constant acceleration result $2as = v_f^2 - v_0^2$

$s = 2.5, \quad a = 3.27, \quad v_0 = 0$

$\Rightarrow v_f = 4.04 \, m/s \qquad \Rightarrow \omega_f = \dfrac{v_f}{r} = 26.9 \, rad/s$

$KE_{rot} = \tfrac{1}{2} I \omega_f^2 = \tfrac{1}{2}(0.18)(26.9)^2 = 65.3 \, J$

7-30

A 7 kg bowling ball, radius 12 cm, travels down a bowling lane at speed 2.0 m/sec, rolling without slipping. Compute these values for the ball:
 (a) Its net kinetic energy.
 (b) The magnitude of its net angular momentum with respect to the base of the pin the ball is about to hit.

***

With respect to the ball's center of mass, kinetic energy (KE) and angular momentum (L) both factor into translational and rotational terms:

$$KE = \tfrac{1}{2} m v_{CM}^2 + \tfrac{1}{2} I_{CM} \omega^2 \quad , \quad L = m v_{CM} r + I_{CM} \omega \quad .$$

For a uniform sphere, $I_{CM} = \tfrac{2}{5} m r^2$; and $\omega = v_{CM}/r$ if there is no slipping.

Thus $KE = \tfrac{1}{2} m (1 + \tfrac{2}{5}) v_{CM}^2 = \tfrac{1}{2} \times 7 \times \tfrac{7}{5} \times 2^2 = \underline{19.6 \text{ J}}$ (a)

$L = m v_{CM} r (1 + \tfrac{2}{5}) = 7 \times 2 \times 0.12 \times \tfrac{7}{5} = \underline{2.35 \text{ kg m}^2/\text{sec}}$ (b)

7-31

A 0.20 kg mass attached to a string is placed in circular motion on a horizontal frictionless surface, at a radial distance of 0.4 m from the center, and with an angular velocity of 6.0 rad/s. The other end of the string passes through a hole in the center of the horizontal surface.

a) What constant force F must be exerted at the end of the string in order to maintain this angular velocity and radius?

b) The string is pulled, by increasing the force F, and the radius is reduced to 0.2 m. What is the angular velocity at the reduced radius?

c) How much work was done in reducing the radius from 0.4 m to 0.2 m?

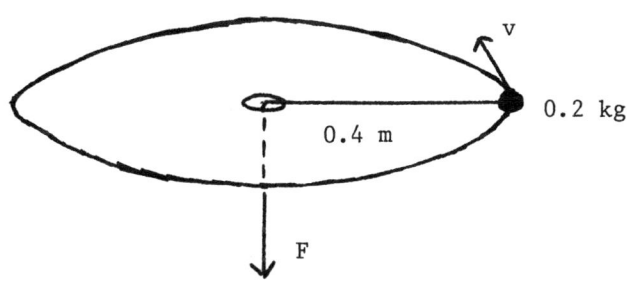

****

a) The force F must provide the required centripetal force. $F_c = ma_c = m\omega^2 r = 0.2\,kg \times (6\,rad/s)^2 \times 0.4\,m = \underline{2.88\,N}$

b) Centripetal force is perpendicular to the tangential velocity, and produces no torque. Thus angular momentum is conserved. $\therefore L_f = L_i$, $L = I\omega$, $I = mr^2$.

$\therefore I_f \omega_f = I_i \omega_i$

$m r_f^2 \omega_f = m r_i^2 \omega_i$; $\omega_f = \omega_i \dfrac{r_i^2}{r_f^2} = 6\,rad/s \dfrac{(0.4\,m)^2}{(0.2\,m)^2}$

$\omega_f = \underline{24\,rad/s}$.

c) Work $= \Delta KE = \tfrac{1}{2} I_f \omega_f^2 - \tfrac{1}{2} I_i \omega_i^2 = \tfrac{1}{2} m (r_f^2 \omega_f^2 - r_i^2 \omega_i^2)$

$= \tfrac{1}{2} \times 0.2\,kg \left((0.2\,m \times 24\,rad/s)^2 - (0.4\,m \times 6\,rad/s)^2\right)$

$= \underline{1.73\,J}$.

Note: one must calculate ΔKE rather than $F \cdot \Delta r$ for the work since the force F is not constant.

7-32

A bowling ball of mass 6 kg and radius .08 m starts from rest rolls without slipping 3 m down an incline which makes an angle of 37° with the horizontal as shown. (a) Use energy methods to find the speed of the ball when it reaches the bottom of the incline. (b) What was the angular acceleration of the ball?

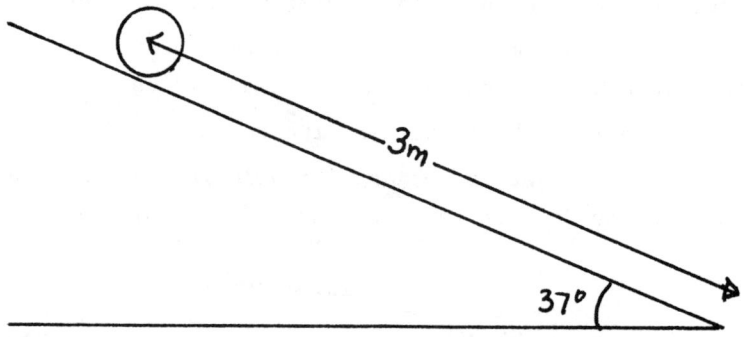

***

a) $\Delta U + \Delta K = 0$

$K_2 - K_1 = -\Delta U = -(U_2 - U_1) = U_1 - U_2 = mgh$

$\left(\frac{1}{2} m V_{cm}^2 + \frac{1}{2} I \omega^2\right) - 0 = mgh$

$\frac{1}{2} m V_{cm}^2 + \frac{1}{2} \left(\frac{2}{5} m R^2\right) \frac{V_{cm}^2}{R^2} = mgh$

$\frac{7}{10} \cancel{m} V_{cm}^2 = \cancel{m} gh$

$V_{cm} = \sqrt{\frac{10}{7} gh} = \sqrt{\frac{10}{7} \frac{(9.8m)}{s^2} 3m \sin 37°}$

$= 5.0 \, m/s$

b) $V_{cm}^2 - V_0^2 = 2 a_{cm} (3m - 0)$

$a_{cm} = \frac{(5.0 m/s)^2 - 0}{2(3m)} = 4.2 \, m/s^2$

$\alpha = \frac{a_{cm}}{R} = \frac{4.2 \, m/s^2}{.08 m} = 52.5 \, rad/s^2$

7-33

A gold hula-hoop, a solid copper cylinder, a solid iron sphere, and a cube of wood enter a pure rolling race. Each starts from rest at the top of an inclined plane which is 19.6 m high and 58 m long. The cube needs some help, so the hoop (H), cylinder (C), and sphere (S) agree to let the cube use a frictionless portion of the inclined plane. Find the speed of each object at the bottom of the inclined plane, neglecting the effects of air friction.

****

For H, C, and S the moments of inertia are MR^2, $\frac{1}{2}MR^2$, and $\frac{2}{5}MR^2$ where M = mass of object and R = radius of object.

Conservation of mechanical energy ⇒

for H: $\frac{1}{2}Mv^2 + \frac{1}{2}(MR^2)\omega^2 = Mgh$
$v^2 + v^2 = 2gh \Rightarrow v^2 = 2gh/2$
$\Rightarrow v = \sqrt{gh} = [9.80 \frac{m}{sec^2} \times 19.6 \, m]^{1/2} = \boxed{13.9 \, m/sec}$

for C: $\frac{1}{2}Mv^2 + \frac{1}{2}(\frac{1}{2}MR^2)\omega^2 = Mgh$
$2v^2 + v^2 = 4gh \Rightarrow v^2 = 4gh/3$
$\Rightarrow v = \sqrt{\frac{4}{3}gh} = [\frac{4}{3}(9.80 \frac{m}{sec^2}) \times 19.6 \, m]^{1/2} = \boxed{16.0 \, m/sec}$

for S: $\frac{1}{2}Mv^2 + \frac{1}{2}(\frac{2}{5}MR^2)\omega^2 = Mgh$
$5v^2 + 2v^2 = 10gh \Rightarrow v^2 = 10gh/7$
$\Rightarrow v = \sqrt{\frac{10}{7}gh} = [\frac{10}{7}(9.80 \frac{m}{sec^2}) \times 19.6 \, m]^{1/2} = \boxed{16.6 \, m/sec}$

for cube: $\frac{1}{2}Mv^2 = Mgh$
$v^2 = 2gh \Rightarrow v^2 = 2gh$
$\Rightarrow v = \sqrt{2gh} = [2(9.80 \frac{m}{sec^2}) \times 19.6 \, m]^{1/2} = \boxed{19.6 \, m/sec}$

The fraction of the kinetic energy which was TRANSLATIONAL differred, so the speeds at the bottom of the inclined plane were different.

7-34

A uniform disk of radius R=20cm and mass M=2.5kg is mounted on an axle supported in fixed frictionless bearings. A cord is wrapped around the rim of the wheel and a falling mass m attached to the cord causes the wheel to rotate starting from rest. The moment of inertia of the disk is ½MR², and the tension in the cord is 5.0N. (a) Find the angular acceleration of the wheel and the tangential acceleration of a point on the rim. (b) Find the mass m attached to the cord. (c) Calculate the kinetic energy of the mass m, the kinetic energy of the pulley and the magnitude of its angular momentum after the mass has fallen a distance of 120 cm, starting from rest.

****

(a) $\sum \tau = I\alpha \qquad TR = I\alpha$
$(5)(.2) = \frac{1}{2}(2.5)(.2)^2 \alpha \qquad \alpha = 20 \text{ rad/sec}^2$
$a_T = (0.2)(20) = 4 \text{ m/sec}^2$

(b) $mg - T = ma$
$m(9.8) - 5 = m(4)$
$m = .862 \text{ kg}$

(c) $K_i + U_i = K_f + U_f \qquad [I = \frac{1}{2}(2.5)(.2)^2 = 0.05 \text{ kg-m}^2]$
$0 + mgh = (\frac{1}{2}mv^2 + \frac{1}{2}I\omega^2) + 0 \qquad \omega = \frac{V}{R} = \frac{V}{0.2}$
$(.862)(9.8)(1.2) = \frac{1}{2}(.862)v^2 + \frac{1}{2}(.05)\frac{v^2}{(.2)^2}$
$V = 3.1 \text{ m/sec}$

$K_{\text{mass m}} = \frac{1}{2}mv^2 = \frac{1}{2}(.862)(3.1)^2 = 4.14 \text{ J}$

$K_{\text{pulley}} = \frac{1}{2}I\omega^2 = \frac{1}{2}(.05)\left(\frac{3.1}{.2}\right)^2 = 6 \text{ J}$

Ang. momentum $= L = I\omega = (0.05)\left(\frac{3.1}{.2}\right) = .775 \frac{\text{kg-m}^2}{\text{sec}}$

7-35

A massless string is wound around a thin cylindrical <u>shell</u> of mass M and radius R. The end of the string is held fixed and the cylindrical shell, initially at rest, is then dropped so that the string unwinds, causing the shell to rotate as shown. After an interval of time, t, the cylindrical shell has fallen a distance h and is moving downward with speed v.

(a) What is the angular speed of rotation, ω, at time t?
(b) What is the total kinetic energy (translational + rotational) of the shell at t?
(c) Show that $v = \sqrt{gh}$.
(d) Find the tension in <u>the</u> string, T.
(e) Show that $t = 2\sqrt{h/g}$.

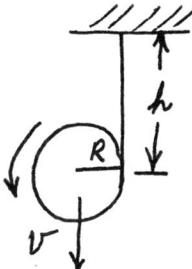

****

(a) $\boxed{\omega = v/R}$

(b) $K = \frac{1}{2}Mv^2 + \frac{1}{2}I\omega^2$; $I = MR^2$
$\Rightarrow K = \frac{1}{2}Mv^2 + \frac{1}{2}MR^2\omega^2 = Mv^2$
$\boxed{K = Mv^2}$

(c) CONSERVATION OF ENERGY $\Rightarrow \Delta K + \Delta U = 0$.
$\Rightarrow K = -\Delta U = Mgh$
$\Rightarrow v^2 = gh$
$\boxed{v = \sqrt{gh}}$

(d) $F = Mg - T = Ma$
$v^2 = 2ah \Rightarrow a = g/2$
$\Rightarrow \boxed{T = Mg/2}$

(e) $v = at \Rightarrow \sqrt{gh} = \frac{g}{2}t$
$\Rightarrow \boxed{t = 2\sqrt{h/g}}$

7-36

A cylindrical drum of radius 0.20 m has a rotational inertia of 1.0 kg·m² about its axis A. A rigid, but light (negligible mass), rod one meter long is rigidly attached to the drum with its center also at A. Point masses of size 3 kg each are attached at each end of the rod. A constant force F is applied to the system by means of a cord wrapped around the drum. Initially (t = 0), the system is at rest. At t = 10 seconds, it has an angular velocity of 4π radians/s. We prefer, but don't require, that you leave any factors of "π" indicated in your answers.

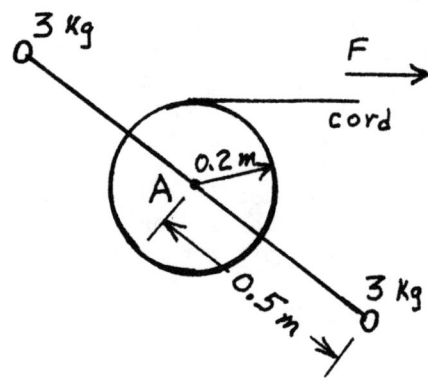

(a) Find the number of revolutions made by the system during the 10 seconds.
(b) Find <u>both</u> components of the <u>linear</u> acceleration of one of the 3 kg masses at t = 10s.
(c) Find the kinetic energy of the system at t = 10s.
(d) Find the size of F.

* *

(a) Constant force implies constant torque, τ, and therefore constant angular acceleration, α, from angular velocity $\omega_0 = 0$ to $\omega = 4\pi$ rad/s in 10 s:

$$\therefore \alpha = (\omega - \omega_0)/t = 0.4\pi \text{ rad/s}^2 \text{ and}$$
$$\Delta\theta = \omega_0 t + \tfrac{1}{2}\alpha t^2 = \tfrac{1}{2}(0.4\pi)10^2 = 20\pi \text{ rad or 10 rev.}$$
$$\{\text{OR } \Delta\theta = \tfrac{1}{2}(\omega + \omega_0)t = \tfrac{1}{2}(4\pi + 0)\cdot 10 = 20\pi \text{ rad} = 10 \text{ rev.}\}$$

(b) The tangential component of acceleration of a point distance r (= 0.5 m) from the axis of a rotating rigid body with angular acceleration α (= 0.4 π) in units rad/s², is
$$a_T = r\alpha = 0.5 \cdot 4\pi = 0.2\pi \text{ m/s}^2, \text{ perpendicular}$$
to the line from point to axis (i.e. perpendicular to rod).

The centripetal component is, if ω (= 4π) is the angular velocity in rad/s
$$a_c = v^2/r = r\omega^2 = 0.5(4\pi)^2 = 8\pi^2 \text{ m/s}^2$$
directed from the point to the axis (i.e. along rod, toward A)

(c) The kinetic energy, K, of rigid rotation is $\tfrac{1}{2}I_A\omega^2$, where the angular measure must be "radian" and I_A is the rotational inertia (or moment of inertia) of the system

$$I_A \text{ (Def.)} = \underset{System}{\sum m_i r_i^2} = \underset{Drum}{\sum m_i r_i^2} + \underset{Rod}{\sum m_i r_i^2} + m_1 r_1^2 + m_2 r_2^2$$
$$= 1.0 \text{ (given)} + 0 \text{ (given)} + 3(0.5)^2 + 3(0.5)^2$$
$$= 2.5 \text{ kg} \cdot m^2$$

When $\omega = 4\pi$ rad/s
$$K = \tfrac{1}{2} I_A \omega^2 = \tfrac{1}{2}(2.5 \text{ kg} \cdot m^2)(4\pi \text{ rad/s})^2 = 20\pi^2 \text{ joules}$$

(d) Two ways to find force F.
   (i) Work-energy theorem: $\Delta K =$ Work done by F
$$\Delta K = K - K_o = 20\pi^2 \text{ J}$$
Work by $F = F \times$ distance F must move
$$= F \times \text{cord length pulled from drum in 10 rev.}$$
$$= F \times (2\pi \cdot \text{radius of drum}) \times 10$$
$$= F \cdot 2\pi \cdot 0.2 \cdot 10 = 4\pi F$$
$$\therefore 20\pi^2 = 4\pi F \Rightarrow F = 5\pi \text{ N}$$

OR (ii) torque of F about $A = I_A \alpha$
   torque of $F = F \times$ lever arm $= 0.2 F$
   from above: $\alpha = 0.4\pi$ rad/s^2, $I_A = 2.5$ kg·m^2
$$\therefore 0.2 F = (2.5)(0.4\pi)$$
$$F = 5\pi \text{ N}$$

7-37

Consider a wheel (radius r = .5m, mass m = 5kg., and moment of inertia I = 6kg.m²) allowed to roll without sliding down an inclined plane as shown. What will its linear velocity be at the bottom of the incline?

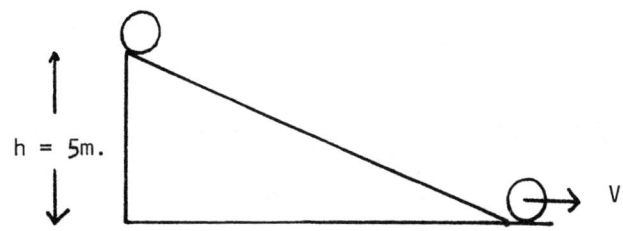

h = 5m.

****

From the cons. of Energy

$$mgh = \tfrac{1}{2} I \omega^2 + \tfrac{1}{2} m v^2$$

But $\omega = \dfrac{v}{r}$

$$\therefore mgh = \tfrac{1}{2} I \dfrac{v^2}{r^2} + \tfrac{1}{2} m v^2$$

$$\therefore \dfrac{mgh}{\tfrac{1}{2}\dfrac{I}{r^2} + \dfrac{m}{2}} = v^2$$

$$\dfrac{5 \cdot 9.8 \cdot 5}{\tfrac{1}{2}\dfrac{6}{(.5)^2} + \tfrac{5}{2}} = v^2$$

$$6.7 \ \dfrac{m^2}{s^2} = v^2$$

$$2.6 \ \dfrac{m}{s} = v$$

7-38

A thin uniform rod is pivoted at the bottom about a fixed horizontal axis perpendicular to the rod. It is allowed to fall under the action of gravity from a vertical position of rest. (a) Find the angular velocity when the rod reaches a horizontal position. (b) Find the angular acceleration. (c) Find the linear acceleration of the end of the rod.

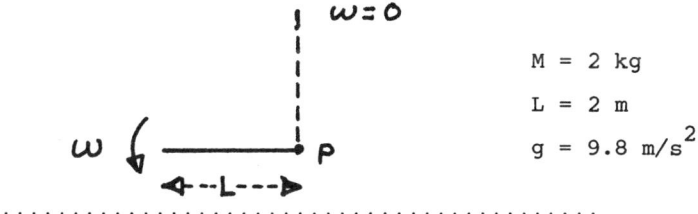

$M = 2$ kg
$L = 2$ m
$g = 9.8$ m/s^2

***

(a) I_P = THE MOMENT OF INERTIA OF THE ROD ABOUT POINT P.

$$I_P = \int r^2 dM = \int_0^L r^2 \lambda \, dr = \frac{\lambda r^3}{3}\Big]_0^L$$

$$I_P = \frac{\lambda L^3}{3} = \frac{M}{L}\frac{L^3}{3} = \frac{1}{3}ML^2$$. NOTE THAT THE

CENTER OF MASS IS AT THE GEOMETRICAL CENTER OF THE UNIFORM ROD, AND THE ROD IS UNDERGOING PURE ROTATION.

<u>ENERGY BEFORE</u> = <u>ENERGY AFTER</u>

$$Mg\left(\frac{L}{2}\right) + \frac{1}{2}I(0)^2 = Mg(0) + \frac{1}{2}I\omega^2$$

$$\omega = \sqrt{\frac{MgL}{I}} = \sqrt{\frac{MgL}{\frac{1}{3}ML^2}} = \sqrt{\frac{3g}{L}} = \sqrt{\frac{3(9.8)}{2}}$$

$$\omega = 3.83 \text{ rad/s}.$$

(b) $\tau = I\alpha$, $\alpha = \frac{\tau}{I} = \frac{Mg\left(\frac{L}{2}\right)}{\frac{1}{3}ML^2} = \frac{3g}{2L} = \frac{3(9.8)}{2(2)}$

$\alpha = 7.35$ rad/s^2 (WHEN HORIZONTAL)

(c) $a = r\alpha$, $a_{END} = (L)\left(\frac{3g}{2L}\right) = \frac{3}{2}g = 14.7$ m/s^2

7-39

A PHONOGRAPH TURNTABLE OF RADIUS=15 Cm IS SPINNING AT 33 REVOLUTIONS PER MINUTE. A LARGE SPIDER IS STANDING AT R=7.5 Cm. HE THEN WALKS TO THE EDGE OF THE RECORD AND STOPS. THE RECORD AND TURNTABLE HAVE A MOMENT OF INERTIA OF .10 KgM^2 SPIDER MASS =500 GRAMS. (ASSUME IT IS A POINT MASS).

(A) FIND NEW ANGULAR VELOCITY IN RPM AND RAD/SEC. (B) FIND THE CHANGE IN KINETIC ENERGY OF TOTAL SYSTEM.

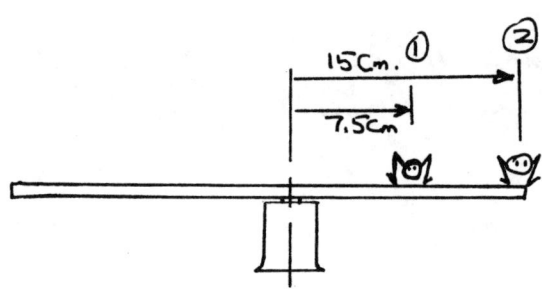

****

Since there is no EXTERNAL TORQUE applied to the system (turntable+spider), the angular momentum is constant.

A) Angular momentum initially equals

$$L_i = \omega_i I_i = 33 \, \text{Rev/min} \, I_i$$

where Ii=initial moment of inertia

$$I_i = I_{TT} + I_{s_i} = .1 \, kg\,m^2 + mR_i^2 = .1\,kg\,m^2 + .5\,kg\,(.075m)^2$$
$$= .103 \, kg\,m^2$$

Angular momentum at end equals

$$L_f = I_f \omega_f$$

where If=final moment of inertia

$$I_F = .1\,kg\,m^2 + .5\,kg\,(.15m)^2 = .111\,kg\,m^2$$

Equating the initial and final angular momentum

$$L_i = L_F \quad \therefore \quad \omega_i I_i = \omega_F I_F$$

Solving for final angular velocity

$$W_F = W_i \frac{I_i}{I_F} = 33 \text{ RPM} * \frac{.103}{.111} = 31 \text{ RPM}$$

or

$$31 \frac{\text{Rev}}{\text{min}} * \frac{2\pi \text{ rad}}{\text{Rev}} \cdot \frac{\text{min}}{60 \text{ sec}} = 3.25 \text{ Rad/sec}$$

B) We cannot assume kinetic energy is constant just because angular momentum is a constant

Using the kintic energy relationship for rotational motion

$$K = \tfrac{1}{2} I \omega^2$$

Initial kinetic energy

$$K_i = \tfrac{1}{2} * .103 \text{ kg m}^2 * \left(33 \frac{\text{Rev}}{\text{min}} * \frac{2\pi}{\text{Rev}} \frac{\text{min}}{60 s}\right)^2 = .615 \text{ J}$$

Final kinetic energy

$$K_F = \tfrac{1}{2} * .111 \text{ kg m}^2 * (3.25 /\text{sec})^2 = .586 \text{ J}$$

Difference in Kinetic energy

$$\Delta K = .586 \text{ J} - .615 \text{ J} = -.029 \text{ J}$$

% Decrease $= \frac{-.029}{.615} = -4.7\%$

7-40

Let us consider the problem of storing energy in a rotating disk such that this stored energy could be used to propel a boat, for instance.
(a) How much energy is stored in a 100 kg disk of radius 30 cm rotating at 20,000 rpm? (b) When considering fuel in the form of gasoline, what volume of gasoline is equivalent, in the form of energy, to that stored in the rotating disk? (Take the energy content of gasoline to be 1.8×10^8 joules/gallon.)

***

(a) The rotational energy is given by,

(1) $$E_{rot} = \tfrac{1}{2} I \omega^2$$

The value of the moment of inertia for a disk is found from

(2) $$I_{disk} = \tfrac{1}{2} m r^2$$

such that

$$I_{disk} = \tfrac{1}{2}(100\,kg)(0.3\,m)^2 = 4.5\,kg\text{-}m^2$$

Now we must convert the angular velocity from having units of revolutions per minute to radians per sec, i.e.

$$\omega = 20{,}000\,\tfrac{rev}{min}\left(\tfrac{1\,min}{60\,sec}\right)\left(\tfrac{2\pi\,rad}{1\,rev}\right) = 2{,}094.395\,\tfrac{rad}{sec}$$

Thus,

$$E_{rot} = \tfrac{1}{2}(4.5\,kg\text{-}m^2)(4{,}386{,}490.843\,rad^2/sec^2)$$

$$= 9{,}869{,}604.397\,joules$$

(b) The energy equivalency in the form of gasoline is thus, from part (a):

$$\frac{E_{rot}}{1.8 \times 10^8\,\tfrac{joules}{gallon}} = \frac{9{,}869{,}604.397\,joules}{1.8 \times 10^8\,\tfrac{joules}{gallon}} = 0.055\,gallon$$

7-41

A uniform metal sphere (m = 0.5 kg and R = 0.02 m) is spinning at 1000 rad/sec about a horizontal axis through its center when it is placed on a horizontal surface with its center of mass stationary. Assume that the ball always maintains contact with the surface as its translational speed increases. The coefficient of kinetic friction between the ball and the surface is equal to 0.2.

How far does the ball travel before the rotational speed matches the translational speed and the ball stops slipping?

****

The friction force, due to the ball slipping (spinning) on the surface, creates a torque that slows the rotation rate and at the same time provides the acceleration on the C of M.

Translation:

$$a = f_k/m = \mu g = .2 \times 10 = 2 \, m/sec$$

$$V_T = at = 2t \quad \text{and} \quad X = \tfrac{1}{2} a t^2$$

$I_{CM} = \tfrac{2}{5} MR^2$

$f_k = \mu mg$

$\tau = f_k \cdot \text{Radius}$

Rotation

$$\alpha = \tau/I = \frac{\mu mg R}{\tfrac{2}{5} MR^2} = \frac{5 \mu g}{2R} = 250 \, sec^{-1}$$

$$\omega_f = \omega_0 - \alpha t = 1000 - 250 t$$

When rotation speed matches translation speed

$$V_T = \omega_f R = (1000 - 250 t) \times .02$$

or

$$V_T = 20 - 5t$$

also have $V_T = 2t$

equate two equations $\Rightarrow 2t = 20 - 5t \Rightarrow t = \tfrac{20}{7} = 2.86 \, sec$

$$\therefore X = \tfrac{1}{2} a t^2 = \tfrac{1}{2} \times 2 \times (2.86)^2 = \underline{8.18 \, m}$$

Note energy is not conserved — friction acts.

7-42

A spherical ball of mass m and radius r rolls without slipping or sliding down an inclined track, starting from rest at an altitude H above the ground. It then goes completely around the vertical loop shown, which has a radius R. Derive an equation for the force exerted on the ball by the track at the top of the loop and compare this with the equation you would have derived for a block sliding without friction down the track.

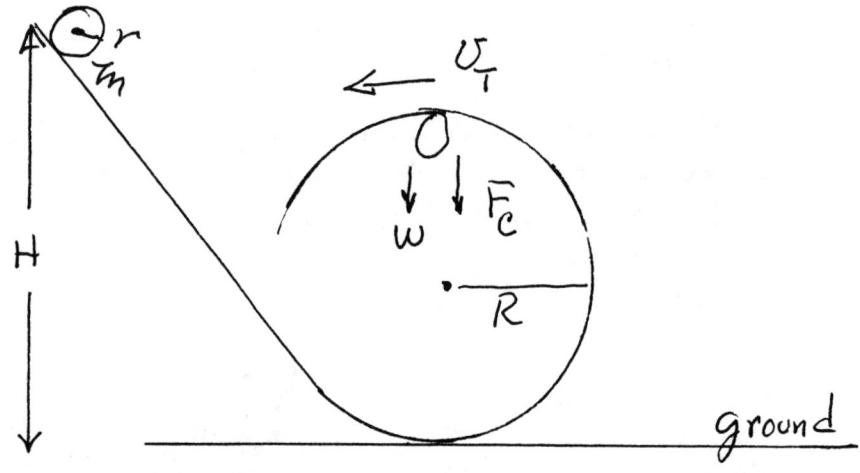

***

The centripetal force at the top is supplied by the weight and the track pushing down on the ball.

$$F_{centripetal} = \frac{mv^2}{R} = mg + F \quad \text{or} \quad ^* \; F = \frac{mv^2}{R} - mg$$

The law of conservation of energy says $\Delta(K.E. + P.E.) = 0$. Note that the friction force which makes the ball rotate does no work since it is a static force (there is no sliding).

Hence $mgH - 2mgR = \frac{1}{2}mv_T^2 + \frac{1}{2}I\omega_T^2$ where $I = \frac{2}{5}mr^2$. Thus

$$^{**} \; mgH - 2mgR = \frac{1}{2}mv_T^2 + \frac{1}{5}mv_T^2 = \frac{7}{10}mv_T^2$$

Substituting mv_T^2 from this equation into * gives

$$F = mg\left[\frac{10H}{7R} - \frac{27}{7}\right]$$

If the ball had been a block sliding without friction down the track we would have gotten

** $mgH - 2mgR = \frac{1}{2}mv_f^2$

OR $F = 2mg\left(\frac{H}{R} - 2.5\right)$

This means that the ball must be released from an altitude $H = 2.7R$ to give $F = 0$, whereas the block need only be released from altitude $H = 2.5$ times R to give $F = 0$ since no rotational energy is involved.

GYROSCOPIC EFFECTS

7-43

A large flywheel is spinning as shown by the arrow on the diagram below. Its axle is fixed rigidly to a boat at points A and B as shown. The coordinates are also fixed to the boat and are centered at the center of the flywheel. (a) What is the direction of the angular momentum of the spinning flywheel? (b) If a force pushes horizontally on the bow of the boat as shown what will be the reaction of the flywheel? (c) If a force pushes vertically from below directed at point B what will be the reaction of the flywheel?

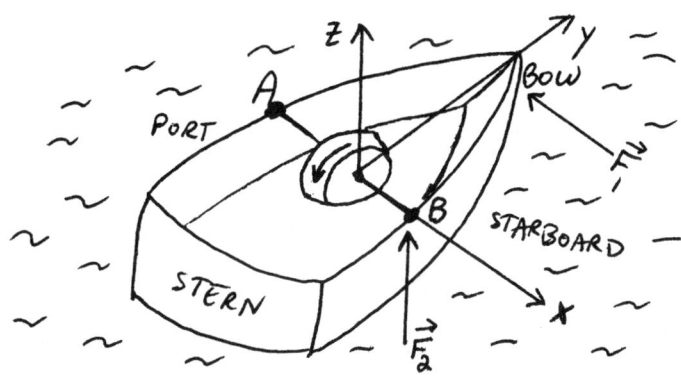

****

(a) Using the Right-Hand-Rule, curl your fingers along the direction of the spinning wheel (shown by the arrow in the diagram) and get the angular momentum ($\vec{L}$) along the $+x$-axis.

(b) The torque ($\vec{\tau}$) is related to $\vec{L}$ by:

$$\vec{\tau} = \frac{d\vec{L}}{dt}$$

The force $\vec{F_1}$ exerted at the bow produces a change in $\vec{L}$ along the $+z$-axis ($\vec{\tau} = \vec{r} \times \vec{F}$ with $\vec{r}$ taken from the center of the flywheel). Thus, the boat will list to PORT, rotating about the y-axis with B moving upward.

(c) The force $\vec{F_2}$ exerted at B produces a change in $\vec{L}$ along the $-y$-axis. Thus, the boat will rotate about the z-axis clockwise as viewed from above.

8
ELASTICITY AND HARMONIC MOTION

STRESS, STRAIN, AND THE ELASTIC MODULI

━━ 8-1

Two round rods, one of steel, the other of brass, are joined end to end. Each rod is 0.6 m long and 2.2 cm in diameter. The combination is subjected to tensile forces of 5,678 N. (a) What is the strain in each rod? (b) What is the elongation of each rod? (c) What is the change in diameter of each rod (excluding edge effects at the point where the two metals are joined)?

MATERIAL	YOUNG'S MODULUS, Y	POISSON'S RATIO, σ
Steel	2.0×10^{11} Pa	0.19
Brass	9.1×10^{10} Pa	0.26

****

(a) Let us first calculate the cross-sectional area of each rod, A: $A = \pi r^2 = \pi d^2/4 = \pi (.022 m)^2 / 4 = 3.801 \times 10^{-4} m^2$

Hence, the stress on each of the rods is

$$\text{stress} = \frac{F_\perp}{A} = \frac{5678 N}{3.801 \times 10^{-4} m^2} = 1.494 \times 10^7 \, Pa$$

Since, by definition,
$$\text{Elastic Modulus} = \text{stress/strain}$$
we may write,

steel strain $= \dfrac{\text{stress}}{Y_{steel}} = \dfrac{1.494 \times 10^7 \text{ Pa}}{2 \times 10^{11} \text{ Pa}} = 7.468 \times 10^{-5}$

brass strain $= \dfrac{\text{stress}}{Y_{brass}} = \dfrac{1.494 \times 10^7 \text{ Pa}}{9.1 \times 10^{10} \text{ Pa}} = 1.642 \times 10^{-4}$

(b) $\Delta l \equiv$ rod elongation and $l_0 \equiv$ initial length of rod

$\Delta l_{steel} = l_0 \text{ (steel strain)} = (0.6 \text{ m})(7.468 \times 10^{-5}) = 4.481 \times 10^{-5} \text{ m}$

$\Delta l_{brass} = l_0 \text{ (brass strain)} = (0.6 \text{ m})(1.642 \times 10^{-4}) = 9.852 \times 10^{-5} \text{ m}$

(c) There is a proportionality between the fractional change in width of the rod and its fractional change in length, i.e.

$\Delta w/w_0 = \Delta d/d_0 = -\sigma \Delta l/l_0$ or $\Delta d = -\sigma d_0 \Delta l/l_0$; $\sigma \equiv$ Poisson's ratio

Thus, the change in diameter in each rod is given as:

$\Delta d_{steel} = \dfrac{-(.19)(.022 \text{ m})(4.481 \times 10^{-5} \text{ m})}{(0.6 \text{ m})} = -3.122 \times 10^{-7} \text{ m}$

$\Delta d_{brass} = \dfrac{-(.26)(.022 \text{ m})(9.852 \times 10^{-5} \text{ m})}{(0.6 \text{ m})} = -9.392 \times 10^{-7} \text{ m}$

The minus sign in each of the values above indicates a decreasing diameter with increasing length for both the brass and steel rods.

8-2

A rod of negligible weight is supported at its ends by wires A and B. The wires are of equal length, and the rod is 4 meters long. The cross-sectional area of wire B is three times the cross-sectional area of wire A, and Young's modulus for both wires is 3.6×10^{11} N/m². At what point along the rod should a 25 N weight be suspended in order to produce equal stresses in wires A and B?

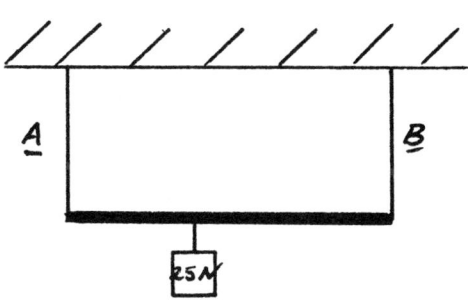

****

① $A_B = 3 A_A$

② equal stresses ⇒ equal force-to-area ratios

⇒ $\dfrac{F_A}{A_A} = \dfrac{F_B}{A_B}$

⇒ $\dfrac{F_A}{F_B} = \dfrac{A_A}{A_B} = \dfrac{1}{3}$ (from ①)

free-body diagram for rod:

$\uparrow F_A \qquad 0 \qquad \uparrow F_B$
$\leftarrow x \rightarrow | \leftarrow 4-x \rightarrow$
$\qquad \downarrow 25N$

from free body diagram: take sum of torques about 0

$\Sigma \tau = 0 = F_A x - F_B(4-x)$

⇒ $\dfrac{F_A}{F_B} \cdot x = (4-x) \Rightarrow \dfrac{x}{3} = 4-x$

⇒ $\boxed{x = 3m}$

(note that the magnitude of the weight and the young's modulus for the wire are irrelevant.)

HOOKE'S LAW AND SIMPLE HARMONIC MOTION

8-3

IN ORDER TO DEMONSTRATE THE FEATURES OF SPRINGS AND HARMONIC MOTION, A PHYSICS PROFESSOR CLAMPED A SPRING TO THE CEILING OF THE PHYSICS LAB. WITH THE AID OF A STEP-LADDER, HE TOOK HOLD OF THE END OF THE SPRING. HOLDING THE SPRING, HE SLOWLY CLIMBED DOWN AND STEPPED OFF THE LADDER, THUS HANGING FREELY IN MID-AIR. THE SPRING EXTENDED 3 METERS FROM ITS ORIGINAL POSITION. NEXT, HE ASKED A STUDENT TO TAKE HOLD OF HIS FOOT, PULL HIM DOWN ANOTHER 2 METERS, PAUSE AND LET GO.

***

A) By Hooke's Law, the force exerted by a spring equals the spring constant times the extension from its equilibrium position. At the professor's new equilibrium position, his weight must equal the force due to the spring.

$$F_s = W \quad \text{where} \quad F_s = ky$$

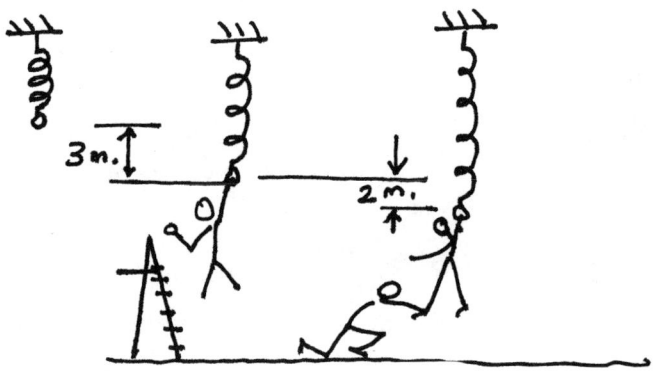

Therefore, the spring constant can be found

$$ky = W$$

$$\text{or} \quad k = W/y = Mg/y = \frac{200 \text{ kg} \times 9.8 \text{ m/s}^2}{3 \text{ m}} = 653 \text{ N/m}$$

B) The frequency is related to the mass and the spring constant

$$f = \frac{\omega}{2\pi} = \frac{1}{2\pi}\sqrt{k/m} = \frac{1}{2\pi}\sqrt{\frac{653 \text{ N/m}}{200 \text{ kg}} \cdot \frac{\text{kg m/s}^2}{\text{N}}} = .29/\text{s}$$

The period is the inverse of the frequency

$$T = 1/f = \frac{1}{.29/sec} = 3.5 \text{ sec}.$$

The equilibrium position of the professor during oscillation is the location he arrived at in part A

i.e... 3 m. below the original spring position

C) The amplitude, or maximum displacement must be the distance to which he was pulled before being released, since his velocity there was zero.

$$\therefore A = 2m.$$

The maximum velocity is found using the relation

$$V_{max} = \omega A = 2m\sqrt{k/m} = 2m\sqrt{\frac{653}{200}} = 3.6 \text{ m/s}$$

The maximum acceleration is found by using the simple relationship

$$a_{max} = \omega^2 A$$
$$= 3.27/s^2 * 2m = 6.5 \text{ m/s}^2$$

8-4

A tiny object near the earth's surface moves vertically in simple harmonic motion with a frequency of 1000 Hz. The object rests on a horizontal platform and would leave the platform if the maximum acceleration of the object became larger than 9.80 (meter/sec)/sec. Use this data to compute the maximum amplitude of vibration which the tiny object can have without leaving the platform, and the time required for this object to move from the lowest to the highest point of its vibration.

****

$\nu = 1000$ Hz

$|a_{max}| = g$

$ma = -ks$ (Hooke's law)

$|m\, a_{max}| = kA$

$A = \left(\dfrac{m}{k}\right)|a_{max}|$

$A = \left(\dfrac{m}{k}\right) g$

$T = 2\pi \sqrt{\dfrac{m}{k}}$

$\dfrac{m}{k} = \dfrac{T^2}{4\pi^2} = \dfrac{1}{4\pi^2 \nu^2}$

$A = \left(\dfrac{1}{4\pi^2 \nu^2}\right) g$

$A = \dfrac{9.80 \text{ m/sec}^2}{4\pi^2 (10^3 \text{ Hz})^2}$

$\boxed{A = 2.48 \times 10^{-7} \text{ meter}}$

$T = \dfrac{1}{\nu} = \dfrac{1}{1000 \text{ Hz}} = 1.000 \times 10^{-3} \dfrac{\text{seconds}}{\text{cycle}}$

$t = \dfrac{T}{2} = \boxed{0.500 \text{ millisecond}}$

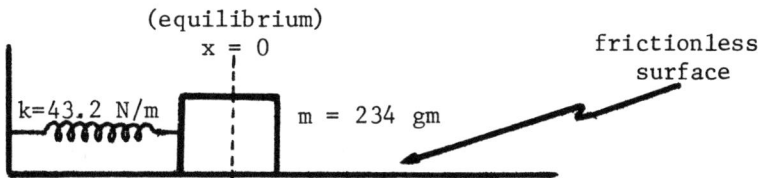

Let us consider the mass above to be characterizing simple harmonic motion when oscillating. Assuming the spring obeys Hooke's law, $\vec{F}_s = -k(\vec{\Delta x})$, find (a) the frequency of oscillation, ν, and the time required for one oscillation about the equilibrium position, T. The amplitude of this simple harmonic motion is 4.32 cm. Let us now consider a particular instant of the oscillating mass, where the mass is located 2.34 cm to the left of the equilibrium position. At this particular location, find (b) the velocity, and (c) the acceleration of the mass, m.

****

(a) From Newton's second law we may write,

(1) $\qquad F_x = ma_x = m\, d^2x/dt^2$

Setting the force of the spring, from Hooke's law, equivalent to expression (1), we have

(2) $\qquad -kx = m\, d^2x/dt^2 \quad \text{or} \quad d^2x/dt^2 = -(k/m)x$

Knowing the mass characterizes simple harmonic motion, its position may be expressed in a sinusoidal form:

(3) $\qquad x(t) = A\cos(\omega t + \varphi);\ \varphi = \text{phase angle and } A = \text{amplitude}$

and

(4) $\qquad dx/dt = -A\omega \sin(\omega t + \varphi)$ and $d^2x/dt^2 = -A\omega^2 \cos(\omega t + \varphi) = -\omega^2 x$

From expressions (2) and (4),

(5) $\qquad k/m = \omega^2 = (2\pi\nu)^2;\ \omega = 2\pi\nu$

Solving (5) for ν yields;

$\nu = (1/2\pi)\sqrt{k/m} = (1/2\pi)\sqrt{(43.2\,N/m)/(.234\,kg)} = 2.162\,Hz$

The period is related to the frequency by

$T = 1/\nu = 1/2.162\,Hz = 0.462\,sec$

(b) To obtain a numerical value for the velocity we must determine the value of $(\omega t + \varphi)$ in expression (3) by substitution known values for A and x.

$x = A\cos(\omega t + \varphi);\ (\omega t + \varphi) = \cos^{-1}(x/A) = \cos^{-1}(-2.34/4.32)$
$\qquad\qquad\qquad\qquad\qquad\qquad = \cos^{-1}(-.542) = 122.79°$

$v = -A\omega \sin(\omega t + \varphi) = -(.0432\,m)(13.584\,Hz)\sin(122.79°) = -0.493\,m/sec$

(c) $a = -\omega^2 x = -(13.584\,\tfrac{rad}{sec})^2 (-0.0234\,m) = 4.318\,m/sec^2$

8-6

A 300 gram mass on a frictionless table is attached to a horizontal spring, the other end of the spring being attached to a wall. The mass is set into vibration, and a timer indicates that it completes three full vibrations in 12 seconds. The closest approach to the wall is 2 cm, and the furthest distance is 16 cm. Determine (a) the amplitude of vibration (b) the spring constant (c) the speed and acceleration of the mass at points 9 cm, 12 cm, and 16 cm from the wall.

***

(a) $A = \frac{1}{2}(16-2) = 7$ cm

(b) $T = 12\text{sec}/3 \text{ vibrations} = 4 \text{ sec/vibration}$

$T = 2\pi\sqrt{m/k}$ $4 = 2\pi\sqrt{.3/k}$ $k = .74$ N/m

(c) The point 9 cm from the wall is the equilibrium point ($x=0$). At 9 cm,

$V = \omega A$ $\omega = \frac{2\pi}{T} = \frac{2\pi}{4} = 1.57$ rad/sec

$V = (1.57)(7) = 11$ cm/sec

$a = -\omega^2 x = 0$

At 12 cm from the wall, $x = 12 - 9 = 3$ cm

$V = \omega\sqrt{A^2 - x^2} = 1.57\sqrt{7^2 - 3^2} = 9.93$ cm/sec

$a = -\omega^2 x = -(1.57)^2(3) = -7.4$ cm/sec^2

The point 16 cm from the wall is the point of maximum amplitude. At this point,

$V = 0$

$a = -\omega^2 A = -(1.57)^2(7) = -17.3$ cm/sec^2

8-7

The graph of the position of a simple harmonic oscillator as a function of time is shown below. The phase constant for this diagram is $\phi = \pi/4$. (a) What is the period of the oscillator? (b) What is the amplitude of the oscillator? (c) Write the equation of the oscillator. (d) What is the speed of the oscillator at t = 6 seconds? (e) On the diagram below, draw a vertical line that represents a new t = 0 axis for a phase of $\phi = \pi$.

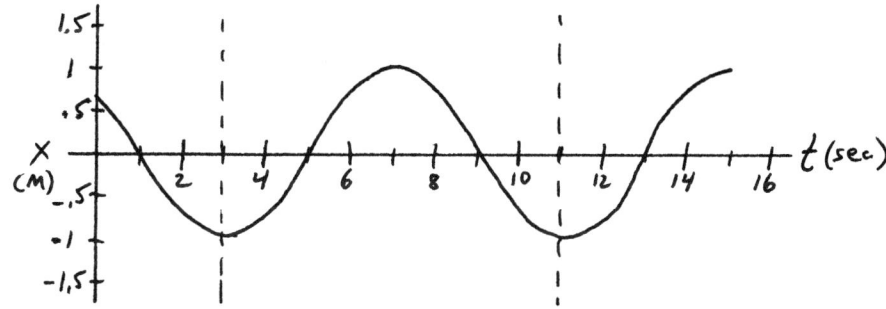

****

(a) $T = 8$ sec ... you must take a complete cycle to find T: beginning when x = 0 at t = 1 sec, it again becomes 0 WITH THE OSCILLATOR MOVING IN THE SAME DIRECTION at t = 9 sec.

(b) The amplitude (A) is defined as the maximum positive displacement for a symmetric oscillator. Therefore, $A = 1$ M.

(c) In general $x = A \cos(\omega t + \phi)$
where $A = 1$ M
$\omega = \frac{2\pi}{T} = \frac{2\pi}{8}$ rad/sec
$\phi = \pi/4$ rad

$x = 1 \cos\left(\frac{2\pi}{8} t + \frac{\pi}{4}\right)$

(d) $v = \frac{dx}{dt} = -\frac{2\pi}{8} \sin\left(\frac{2\pi}{8} t + \frac{\pi}{4}\right) = +0.56$ m/sec when t = 6 sec

(e) when $\phi = \pi \rightarrow x = A \cos(\omega t + \pi) = -A \cos \omega t$ is a pure negative cosine curve. Thus, a vertical axis at t = 3 sec or at t = 11 sec will be correct.

8-8

Two masses of 4 and 7 kg are attached to a vertical spring. As a result the spring stretches 0.15m from its unstretched length. At t=0, the string connecting the 4 kg mass is cut. For the resulting simple harmonic motion of the 7 kg block calculate the following.
a) The frequency of the simple harmonic motion.
b) The amplitude of the simple harmonic motion.
c) The maximum and minimum amounts that the spring is stretched.
d) Write an equation for the amount the spring is stretched as a function of time after t=0.

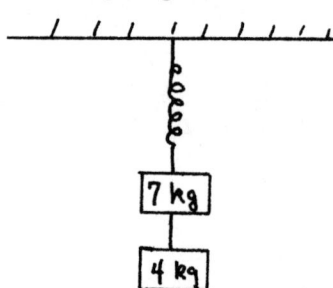

***

a) First calculate the spring constant
$$F = kx \quad \text{for} \quad F = 11(9.8)\,N,\ x = 0.15\,m$$
$$(11)(9.8) = k(0.15) \Rightarrow k = 719\ N/m$$
The oscillating mass is 7 kg

Thus: $\omega = \sqrt{\dfrac{k}{m}} = \sqrt{\dfrac{719}{7}} = 10.1\ rad/s$

b) The oscillations occur about the equilibrium pt.
For a 7 kg load $\Delta L = \dfrac{7(9.81)}{719} = 0.0954\,m$.

The initial displacement is 0.1500 m and at rest.
$\Rightarrow$ Amplitude $= 0.1500 - 0.0954 = 0.0546$

c) $\Delta L_{max} = 0.1500\,m$

$\Delta L_{min} = 0.0954 - 0.0546 = 0.0409\,m$

d) $\Delta L = 0.0954 + 0.0546 \cos(10.1\,t)\ m$

where t is in sec.

8-9

A 2 kg block traveling at 10 m/sec on a horizontal, frictionless track strikes and becomes fastened to the end of an ideal spring without a loss of energy. The other end of the spring is fixed. If the spring is compressed 10 cm before the block stops, what is the period of the resultant harmonic motion?

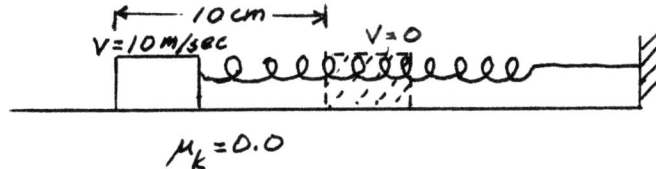

****

The period $T = \frac{1}{f}$; $2\pi f = \omega$; $\omega = \sqrt{\frac{k}{m}}$

To solve problem one must get the spring const, k

From conservation of energy:

Kinetic energy of mass = Potential energy Spring

$$\frac{1}{2} m v^2 = \frac{1}{2} k x^2$$

or

$$k = \frac{mv^2}{x^2} = \frac{2 kg \times (10 \text{ m/sec})^2}{(.1 m)^2} = 20000 \text{ N/m}$$

∴ $\omega = \sqrt{\frac{k}{m}} = \sqrt{\frac{20000}{2}} = 100 \text{ sec}^{-1}$

and

$$T = \frac{2\pi}{\omega} = \frac{6.28}{100} = \underline{.0628 \text{ sec}}$$

8-10

Given: a mass m, the displacement from equilibrium satisfying the relation

$$x = 3.0 \sin(4t + 2),$$

where x is displacement from equilibrium in meters and t is time in seconds. Take m = 0.30 kg.

(a) What is the amplitude?
(b) What is the value of angular speed?
(c) What is the value of the initial phase angle?
(d) What is the period of the oscillation?
(e) What is the value of the frequency?
(f) Find the maximum speed of the mass m.
(g) Find the total mechanical energy of mass m.
(h) Given the expression for x(t) stated at the outset, find the expression for the acceleration as a function of time, a(t).
(i) From the result of (h) write the expression for F(t), the force applied to the particle.
(j) By comparing the result found in (i) with the expression for x(t) given at the outset, show that the force you've obtained is a linear, restoring force.

****

(a) $A = 3.0$ m

(b) $\omega = 4.0 \frac{(rad.)}{s}$

(c) $\theta_0 = 2$ (radians)

We have the form $x = A \sin(\omega t + \theta_0)$.

(d) $T = \frac{1}{f}$; $2\pi f = \omega$, $\frac{1}{f} = \frac{2\pi}{\omega}$. $\therefore T = \frac{2\pi}{\omega} = \frac{2\pi}{4} = \frac{\pi}{2}$ (s)

(e) $f = \frac{\omega}{2\pi} = \frac{4}{2\pi} = \frac{2}{\pi}$ (s^{-1}). Note that $f = \frac{1}{T}$ checks (compare (d) and (e) results)

(f) $v = \frac{dx}{dt} = \frac{d}{dt}(3.0 \sin(4t+2)) = 12 \cos(4t+2)$. $\therefore v_{max} = 12 \frac{m}{s}$.

(g) $E = K + U = K_{max} = U_{max}$.

$K_{max} = \frac{1}{2} m v_{max}^2 = 21.6$ Joule

(h) $a(t) = \frac{d}{dt}(v(t)) = \frac{d}{dt}(12\cos(4t+2)) = -48\sin(4t+2)$ m/s²

(i) $F = ma = 0.30(-48\sin(4t+2)) = -14\sin(4t+2)$ N

(j) From (i), $F = -14\sin(4t+2)$ N.
Also $x = 3.0\sin(4t+2)$ m.
$\therefore F = -\frac{14}{3.0}x = -4.8x$

Note that $F \propto x$ (linear force); and F is opposed, in sense, to the displacement x (as indicated by minus sign), i.e., <u>restoring</u> force.

Note that we have found in the last expression the "spring force", 4.8 N/m.

8-11

A simple harmonic motion is given by the equation y = 3.6 cos(0.8t + 0.24). Find the spring constant k in nt/m if the mass of the oscillating object is 10 grams. The units for y and t are meter and second, respectively.

****

The velocity of the mass v is given by
$$\dot{y} = -2.88 \sin(0.8t + 0.24),$$
and its acceleration a is
$$\ddot{y} = -2.30 \cos(0.8t + 0.24).$$
Now $F = -ky = ma$. Therefore,
$$k = \frac{m\{-2.30\cos(0.8t+0.24)\}}{-3.6\cos(0.8t+0.24)} = (0.01)\left(\frac{2.30}{3.6}\right)$$
$$\cong 6.4 \times 10^{-3} \text{ nt/m} \quad \leftarrow$$

8-12

A 5 kg mass, sitting on a frictionless horizontal surface, is attached to a spring. The system is displaced 8 cm from the equilibrium poistion and released. The period of vibration of the mass is 4 seconds.

a) Find the time required for the mass to travel 4 cm after it is released.

b) Find the force constant of the spring.

****

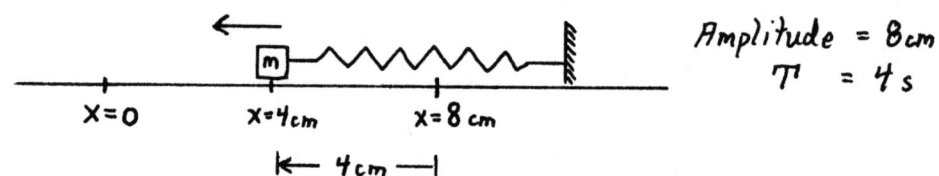

Amplitude = 8 cm
$T = 4s$

a) find displacement equation and solve for t

$$x = A \cos 2\pi f t = A \cos 2\pi \frac{1}{T} t$$

$$x = 8 \cos \frac{2\pi}{4} t \quad cm$$

when $x = 4$ cm

$$4 = 8 \cos \frac{\pi}{2} t \quad \text{or} \quad t = \frac{2}{\pi}\left(\cos^{-1}\frac{1}{2}\right) = .667 s$$

b) $T = 2\pi \sqrt{\frac{m}{K}}$

$$T^2 = 4\pi^2 \frac{m}{K} \quad \text{OR} \quad K = \frac{4\pi^2 m}{T^2} = \frac{4\pi^2 (5)}{16} = 12.3 \frac{N}{m}$$

8-13

A 3 kg block is connected to a spring and is oscillating vertically with simple harmonic motion of 23 cm amplitude and 1.75 seconds period.
(a) What is the spring constant?
(b) What is the maximum velocity of the block?
(c) What is the maximum acceleration of the block?
(d) What is the velocity of the block when it is 8 cm below the highest point reached?
(e) What is the acceleration of the block when it is 8 cm below the highest point reached?

****

(a) $T = 2\pi\sqrt{\dfrac{m}{k}}$ $k = 4\pi^2 \dfrac{m}{T^2} = 4\pi^2 \dfrac{3}{1.75^2} =$ 38.7 N/m

(b) $V_m = \omega A = \dfrac{2\pi}{T} A = \dfrac{2\pi}{1.75}(23) =$ 82.6 cm/s

(c) $a_m = \omega^2 A = \dfrac{4\pi^2}{T^2} A = \dfrac{4\pi^2}{1.75^2}(23) =$ 296 cm/s²

(d) $v = \pm \omega \sqrt{A^2 - x^2} = \pm 3.59\sqrt{23^2 - 15^2} =$ ±62.6 cm/s

(e) $a = -\omega^2 x = -3.59^2 (+15) =$ −193 cm/s²

ENERGY OBSERVATIONS WITH SIMPLE HARMONIC MOTION

8-14

A 1.2 kg block hangs in equilibrium from a spring of constant 3.0 n/m. If it is struck a blow so that it initially travels upward at 0.25 m/s, calculate: The period, the amplitude, and using the cosine representation, write an expression for x (t).

****

a) $T = 2\pi \sqrt{m/k} = 3.97$ sec.

b) using energy principles
$$\tfrac{1}{2} k A^2 = \tfrac{1}{2} m v^2 + \tfrac{1}{2} k x^2$$
$$\tfrac{1}{2} \times 3 \times A^2 = \tfrac{1}{2} \times 1.2 \times (0.25)^2$$
yields
$$A = 0.316 \text{ m}$$

c) $x(t) = A \cos(2\pi t/T + \phi)$

$x(0) = 0 = A \cos(\phi)$

$\phi = \pi/2$

$x(t) = 0.316 \cos(1.58 t + \pi/2)$

8-15

An unloaded spring of force constant 49 N/m hangs as shown in figure A. A 0.5 kg mass is attached and is pulled down 0.15 m below the unloaded position, and is released as shown in figure B. What maximum height measured from the release position does the 0.5 kg mass reach?

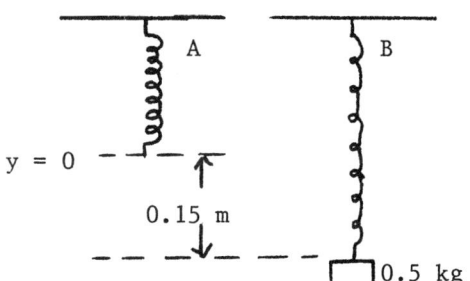

****

Take the unloaded position as $y=0$ as a convenient reference for both PE terms (gravitational and elastic)

$$E_{total} = PE + KE \qquad PE = mgy + \tfrac{1}{2}ky^2 \qquad KE = \tfrac{1}{2}mv^2$$

At any coordinate y, $E_{tot} = mgy + \tfrac{1}{2}ky^2 + \tfrac{1}{2}mv^2$

$v = 0$ at maximum and minimum heights.

The minimum height occurs at $y = -0.15 m$

$$\therefore E_{tot} = mgy + \tfrac{1}{2}ky^2 = mg(-0.15m) + \tfrac{1}{2}k(-0.15m)^2$$

y being the unknown maximum height.

$$0.5\,kg \times 9.8\tfrac{m}{s^2}\, y + \tfrac{49\,N/m}{2}\,y^2 = -0.5\,kg \times 9.8\tfrac{m}{s^2} \times 0.15m + \tfrac{49\,N/m}{2}(0.15m)^2$$

$$24.5\,\tfrac{N}{m}\,y^2 + 4.9N\,y + 0.184\,N\cdot m = 0$$

$$y = \frac{-4.9N \pm \sqrt{(4.9N)^2 - 4 \times 24.5\,N/m \times 0.184\,N\cdot m}}{2 \times 24.5\,N/m}$$

$$y = \frac{-4.9N \pm 2.44N}{49\,N/m} = -0.05m,\ -0.15m$$

The first root is the maximum height, the second is the minimum position (release point). The maximum height above the release position is thus $y - y_0$ or $-0.05m - (-0.15m) = \underline{0.10\,m}$

8-16

A mass m is attached to a spring and executes simple harmonic motion. When the string is stretched the most,

the speed is: largest   zero   neither
and the acceleration is: largest   zero   neither
and the potential energy is: largest   zero   neither
and the kinetic energy is: largest   zero   neither
and the total energy is: largest   zero   neither.

Circle one choice for each, and explain why.

****

Speed is <u>zero</u>; intuitively, it's clear; mathematically, it's because x and v are 90° out of phase: if $x = A\sin\omega t$, then $v_x = \omega A \cos\omega t$.

Acceleration is <u>largest</u>: $|a| = |F/m| = \frac{k}{m}|x|$: proportional to x.

Potential energy is <u>largest</u>: $U = \frac{1}{2}kx^2$

Kinetic energy is <u>zero</u>, because speed is zero and $K = \frac{1}{2}mv^2$.

The total energy is <u>neither</u> zero nor "largest", because it doesn't change with time; it's constant. (It's also $= U = \frac{1}{2}kx^2$ at this particular instant.)

8-17

A mass of 0.5 kg traveling upward with a speed of 10 m/sec strikes and becomes hooked to the end of a hanging ideal spring with negligible loss of energy. If the spring is compressed 20 cm before the mass is brought to rest, what is the amplitude of the resultant harmonic motion? (Do this problem using energy considerations).

****

Note: the point at which the mass becomes connected to the spring is not the equilibrium point for the mass hanging on the spring, therefore the amplitude is not 20 cm. The amplitude of the oscillation is one half the distance between stopping points of the spring. These points may be obtained from the energy equation.

Since the mass has gravitational potential energy, the original point of attachment shall be the reference.

E = Kinetic Energy of mass
$= \frac{1}{2} m v^2 = .5 \times .5 \times 10^2 = 25$ J

Energy at top $= E = \frac{1}{2} k Y_T^2 + mgy$
can be used to get k
$25 = \frac{1}{2} \times .04 \times k + .5 \times 10 \times .2$
$\Rightarrow k = 1200$ N/m

Energy at bottom $= E = \frac{1}{2} k Y_B^2 + mg Y_B$

$25 = 600 Y_B^2 + 5 Y_B$

or

$600 Y_B^2 + 5 Y_B - 25 = 0$

Solving for Y_B one obtains $-.208$ m and $Y_B = +.2$
The second solution is the compression distance.

The amplitude $= \frac{1}{2}(.2 + .208) = .204$ m $= \underline{20.4 \text{ cm}}$

8-18

A dust particle of mass 2×10^{-8} kg sticks to one end of a tuning fork which executes simple harmonic motion of frequency 200 hertz (or cycles/sec) and amplitude 0.6 mm. Preferably, leave factors of "π" indicated.

(a) Let x be the displacement of the particle from its equilibrium, and assume maximum displacement at t = 0. Write an equation for x as a function of time. Put in the correct numerical values for all constants.

(b) Find the largest force needed during the motion to keep the dust particle sticking to the fork.

(c) Find the kinetic energy of the dust particle when it is 0.3 mm from its equilibrium position.

* *

SHM occurs when the acceleration, a, of a mass m is proportional but opposite to its displacement x

(1) $\quad a = d^2x/dt^2 = -\omega^2 x \quad$ [Kinematic definition]

where the constant ω^2 determines the frequency (see below).

OR: SHM occurs when the resultant force F on m is proportional but opposite to x

(2) $\quad F = -kx \quad$ [Dynamic definition]

where the force constant, k, is determined by the forces on m.

BECAUSE $F = ma$, these are equivalent when
$$-kx = m(-\omega^2 x) \quad \text{or} \quad k = m\omega^2$$

The solutions to (1) are $\quad x = A\cos(\omega t + \phi) \quad$ (3)

where $A \equiv$ Amplitude = max. disp. of m, and $\phi \equiv$ phase constant, must fit the particular situation. Since the cosine has period 2π, SHM repeats after a time $T \equiv$ the period (the reciprocal of the frequency, ν) such that

(4) $\quad \omega = 2\pi/T = 2\pi\nu \quad$ (called "angular frequency")

(a) $A = 0.6$ mm; $\nu = 200/s$ so $\omega = 400\pi$ rad/s

At $t=0$, Eq (3) is $x = A\cos(0+\phi) = A$ (given)

$\therefore \cos\phi = A/A = 1 \quad \phi = 0$

Put these constants in (3) to get $x(t)$

(5) $\quad x = (0.6 \text{ mm}) \cos(400\pi t)$

$\quad = 6 \times 10^{-4} \cos(400\pi t)$ in meters

(b) Differentiate Eq (5) once to get
$$v = dx/dt = -0.24\pi \sin(400\pi t) \text{ m/s}$$
and again to get
$$a = dv/dt = -96\pi^2 \cos(400\pi t) \text{ m/s}^2$$

Largest $F = m \times$ largest $a = 2 \times 10^{-8} \text{ kg} \cdot 96\pi^2 \text{ m/s}^2$
$$= 1.92\pi^2 \times 10^{-6} \text{ N}$$

(c) Since kinetic energy, $K = \frac{1}{2}mv^2$ and (from (b)) $v = -0.24\pi \sin(400\pi t)$, we need only find the $\sin(400\pi t)$ when $x = 0.3$ mm:
$$x = 0.6 \cos(400\pi t) = 0.3 \Rightarrow \cos(400\pi t) = \frac{0.3}{0.6} = \frac{1}{2}$$

But $\sin\theta = \sqrt{1-\cos^2\theta}$ so $\sin(400\pi t) = \sqrt{1-\frac{1}{4}} = 0.866$; now $v = -0.24\pi(0.866) = -0.208\pi$ m/s
and $K = \frac{1}{2}(2 \times 10^{-8} \text{ kg})(-0.208\pi \text{ m/s})^2 = 4.3\pi^2 \times 10^{-10}$ J

{On a calculator, simply find $\sin(400\pi t)$ as $\sin[\cos^{-1} 0.5]$.}

8-19

A mass vibrates on a spring with simple harmonic motion. An experimenter then pulls the mass out further and releases it, causing a vibration of double the previous amplitude. Find the ratio of:
(a) the new total energy to the old total energy
(b) the new maximum velocity to the old maximum velocity
(c) the new maximum acceleration to the old maximum acceleration
(d) the new period to the old period.

****

(a) $E_1 = \frac{1}{2} k A_1^2 \qquad E_2 = \frac{1}{2} k (2A_1)^2 = 4 \left[\frac{1}{2} k A_1^2 \right]$
$= 4 E_1$

So $E_2 : E_1 = 4 : 1$

(b) $V_{max} = \omega A \qquad \omega = \sqrt{k/m}$ and is unaffected by change in amplitude.

$\dfrac{V_{2\,MAX}}{V_{1\,MAX}} = \dfrac{A_2}{A_1} = 2 : 1$

(c) $a_{MAX} = -\omega^2 A \qquad \dfrac{a_{2\,MAX}}{a_{1\,MAX}} = 2 : 1$

(d) $T = 2\pi/\omega$ and is therefore unaffected by change in amplitude.

$T_2 : T_1 = 1 : 1$

DAMPED HARMONIC MOTION

8-20

A 200 gram mass is suspended from a spring of strength 40 N/m. The mass is pulled down through 3 cms and then released from rest. Calculate the natural frequency of the motion, and write an expression for the displacement from equilibrium of the mass. Calculate the energy of the motion.
Now suppose the mass and spring are placed in a liquid, so there is a small viscous force acting, whose magnitude is 0.2 N-s/m times the velocity. Calculate the loss of mechanical energy during one cycle of the motion. You may assume the damping is light.

****

Let the downward vertical direction be the y-axis.
Eqn. of motion: $m\ddot{y} + ky = 0 \Rightarrow \omega = \sqrt{\frac{k}{m}} = \sqrt{\frac{40 \text{ N/m}}{.2 \text{ kg}}}$

$\therefore \omega = 14.14$ rads/s

Displacement: $y = 0.03 \cos 14.14t$ m

Energy: $E = \frac{1}{2}mv^2 + \frac{1}{2}ky^2 = \frac{1}{2}ky_{max}^2$

$\therefore E = \frac{1}{2}(40 \frac{N}{m})(.03 m)^2 = 0.018$ J

With damping, eqn. of motion: $m\ddot{y} + ky + b\dot{y} = 0$
$\qquad\qquad\qquad\qquad\qquad\qquad\qquad\qquad$ ↑ .2 N-s/m

$|\text{Power}| = \vec{b\dot{y}} \cdot \vec{\dot{y}} = b\dot{y}^2$

loss of energy per cycle $\Delta E = \int_0^T b\dot{y}^2 dt = bT\langle\dot{y}^2\rangle$

But $\frac{1}{2}m\langle\dot{y}^2\rangle =$ Kinetic energy $= \frac{1}{2}E_{TOTAL}$

$\therefore \langle\dot{y}^2\rangle = \frac{E}{m} \qquad \therefore \Delta E = \frac{bTE}{m} = \frac{bE}{m} \cdot \frac{2\pi}{\omega}$

$\therefore \Delta E = \frac{(.2 \frac{N-s}{m})(.018 \text{ J}) 2\pi}{(.2 \text{ kg})(14.14 \text{ rads/s})} = .008$ J

loss of energy per cycle

8-21

A damped oscillator loses mechanical energy exponentially:
$$E(t) = E_0 \exp(-t/t_c),$$
where t_c is the time constant. In 10 cycles of oscillation, one half of the initial mechanical energy is lost.
(a) What fraction of the energy is lost in each cycle?
(b) If the period of oscillation is 1.0 sec, what is the value of t_c?

****

(a) Let $t = 10T$: $\dfrac{E(10T)}{E_0} = \dfrac{1}{2} = \exp(-10T/t_c) \Rightarrow \dfrac{T}{t_c} = \dfrac{\ln 0.5}{-10} = 0.0693$

Thus, in one period, $\dfrac{E(T)}{E_0} = \exp(-T/t_c) = \exp(-0.0693) = 0.933 = 93.3\%$.

The loss in one period is then $\underline{6.7\%}$.

(b) $t_c = \dfrac{T}{0.0693} = \dfrac{1 \text{ sec}}{0.0693} = \underline{14.4 \text{ sec}}$.

THE SIMPLE PENDULUM

8-22

Consider the possibility that the universal gravitational field is decreasing such that the acceleration due to gravity is decreasing. Using the period of a simple pendulum to measure this change, show that a 1% change in "g" yields a 1/2 % change in the period of the simple pendulum.

****

$T = 2\pi \sqrt{\dfrac{L}{g}}$ $\dfrac{dT}{dg} = 2\pi L^{1/2}\left(-\dfrac{1}{2}g^{-3/2}\right) = -2\pi \dfrac{L^{1/2}}{g^{1/2}} \dfrac{1}{2g}$

$\qquad\qquad\qquad\qquad\qquad = -T \dfrac{1}{2g}$

$\therefore \dfrac{dT}{T} = -\dfrac{dg}{2g}$ $\therefore$ if $\dfrac{dg}{g} = .01$ i.e. a 1% change

$\qquad\qquad\qquad\qquad \dfrac{dT}{T} = -.005$ i.e. a 1/2 % change

8-23

A SOLID SPHERE OF MASS = 2.0 Kg AND DIAMETER = 1.0 Meter IS ATTACHED BY A MASSLESS STRING OF LENGTH = A, WHICH IS ADJUSTABLE. TO WHAT LENGTH SHOULD THE LENGTH, A, BE SET IN ORDER THAT THE PENDULUM COMPLETES 3000 CYCLES EVERY 4.0 HOURS?

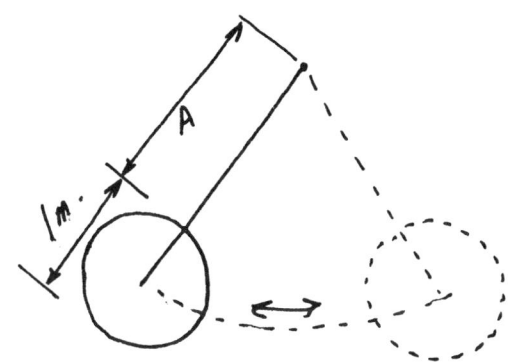

***

The simple pendulum relates the period to the length

$$T = 2\pi\sqrt{L/g}$$

The period is the time for the pendulum to complete one complete cycle. Dividing the total time by the number of cycles, therefore, gives the period.

$$T = 4\text{HRS}/3000 = .0013 \text{ HRS} \times 3600\text{s}/\text{HR} = 4.79 \text{ sec}$$

But L in the pendulum formula is the distance from the center of motion to the center of mass of the sphere. For a sphere, the center of mass is at the center

$$L = A + \tfrac{1}{2}\text{m}.$$

$$\& \quad T = 2\pi\sqrt{(A+.5\text{m})/g}$$

Solving for A, the length of the string

$$g\left(T/2\pi\right)^2 = A + .5\text{m}.$$

$$9.8 \text{ m/s}^2 \left(\frac{4.79 \text{ sec}}{2\pi}\right)^2 - .5 \text{ m}. = A$$

$$A = 5.2 \text{ m}.$$

8-24

A simple pendulum which is 3 meters long has a frequency of 2.0 rad/s when mounted on an elevator which falls with uniform acceleration and it has a frequency of 2.4 rad/s when the elevator has upward acceleration of the same magnitude. What is the magnitude of acceleration of the elevator?

****

Describing problem:

$\ell = 3 m$

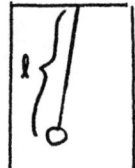

$\omega_d = 2.0$ rad/s

$\omega_u = 2.4$ rad/s

Setting up eqs.

$$\omega_u = \sqrt{\frac{g+a}{\ell}} \quad ; \quad \omega_d = \sqrt{\frac{g-a}{\ell}}$$

$$\Rightarrow \omega_u^2 - \omega_d^2 = \frac{g+a}{\ell} - \frac{g-a}{\ell}$$

$$= \frac{2a}{\ell}$$

$$\Rightarrow a = (\omega_u^2 - \omega_d^2)\ell/2$$

Substituting in numbers:

$$a = [(2.4)^2 - (2.0)^2]\, 3/2$$

$$= 2.64 \text{ m/s}^2$$

8-25

A simple pendulum has a period of 12 seconds. It is displaced slightly from equilibrium and then released from rest. Compute the time required for the bob to return to the equilibrium point. Then compute the time required for the bob to move only half the distance back toward the equilibrium point.

***

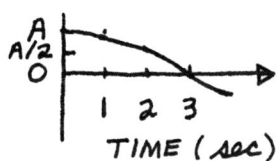
DISPLACEMENT FROM EQUILIBRIUM

To return to equilibrium,

$$t = T/4 = \frac{12 \text{ sec}}{4} = \boxed{3 \text{ seconds}}$$

To move half the distance back to the equilibrium point,

$$x = A \cos \omega t \quad \text{where } x = A/2$$

$$\frac{A}{2} = A \cos \omega t$$

$$\frac{1}{2} = \cos \omega t, \quad \omega t = 60° = \frac{\pi}{3} \text{ radians}$$

$$t = \frac{\pi/3}{\omega} = \frac{\pi/3}{2\pi \nu} = \frac{1}{6\nu} = \frac{T}{6} = \frac{12 \text{ sec}}{6} = \boxed{2 \text{ seconds}}$$

8-26

A simple pendulum of length 0.5 meters is set in motion 3° away from the vertical (as shown). At the bottom of its swing the pendulum's motion is interrupted by an obstruction 0.25 meters long. How long does it take this pendulum to complete one cycle?

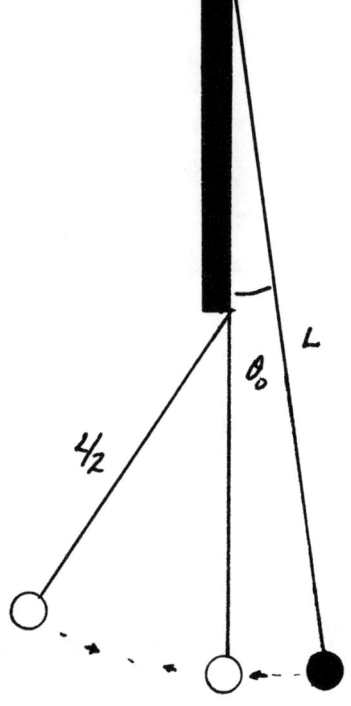

****

The pendulum oscillates for half its cycle as if it had a length L; and for the other half, as if it had a length $L/2$.

$$\Rightarrow \tau_L = 2\pi/\omega_L = 2\pi\sqrt{L/g} \;;\; \tau_{L/2} = 2\pi/\omega_{L/2} = 2\pi\sqrt{\frac{L/2}{g}}$$

$$\tau = \tfrac{1}{2}\tau_L + \tfrac{1}{2}\tau_{L/2} = \tfrac{1}{2}\left(2\pi\sqrt{\frac{.5\,m}{9.8\,m/s^2}} + 2\pi\sqrt{\frac{.25\,m}{9.8\,m/s^2}}\right)$$

$$\Rightarrow \boxed{\tau = 1.21\ sec}$$

THE PHYSICAL PENDULUM

8-27

A metre stick is suspended on a horizontal nail passing through a hole at the 38.0-cm mark so that the stick can execute approximate simple harmonic motion in a vertical plane.
(a) What is its period?
(b) What is the length of a simple pendulum having the same period?

***

(a) The period of a physical pendulum is $T = 2\pi \sqrt{\frac{I}{mgd}}$, where I is the rotational inertia about the point of suspension and can be found using the parallel axis theorem:

$$I = I_c + md^2 = \tfrac{1}{12} m\ell^2 + md^2$$

$$I = m\left[\frac{(1.00\,m)^2}{12} + (.12\,m)^2\right] = (0.0977\,m^2)\,m$$

(↑ metre, ↑ mass)

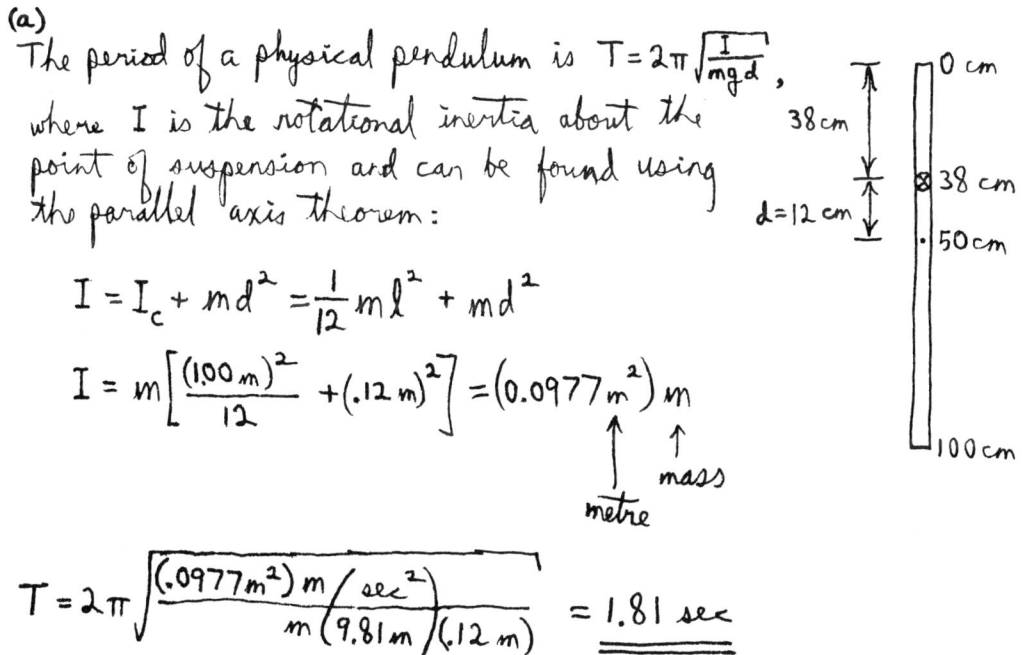

$$T = 2\pi \sqrt{\frac{(.0977\,m^2)\,m\,(sec^2)}{m\,(9.81\,m)(.12\,m)}} = \underline{\underline{1.81\,sec}}$$

(b) $T = 2\pi\sqrt{\frac{\ell}{g}} \Rightarrow \ell = \frac{gT^2}{4\pi^2} = \left(\frac{9.81\,m}{sec^2}\right)\frac{(1.81\,sec)^2}{4\pi^2} = 0.814\,m = \underline{\underline{81.4\,cm}}$

8-28

A physical pendulum has the form of a thin uniform ring pivoted at the top about a fixed axis perpendicular to the plane of the ring. (a) Find the moment of inertia about the axis. (b) Find the period of small amplitude oscillations. (c) Find the center of oscillation and the center of percussion.

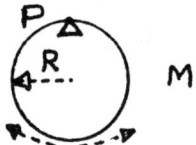

***

(a) The moment of inertia about a parallel axis through the center of mass of the ring is $I_c = \int r^2 dM = MR^2$.

Use the parallel axis theorem:
$I_p = I_c + MR^2$, $I_p = MR^2 + MR^2 = \boxed{2MR^2}$

(b) $T = 2\pi\sqrt{\dfrac{I_p}{Mgr}}$, $T = 2\pi\sqrt{\dfrac{2MR^2}{MgR}} = \boxed{2\pi\sqrt{\dfrac{2R}{g}}}$

Note: A simple pendulum period $= 2\pi\sqrt{\dfrac{L}{g}}$

(c) The center of oscillation is a point, located on a straight line through the pivot P and the center of mass, a distance L below the pivot, where L is the length of a simple pendulum with the same period as the physical pendulum.

> The center of oscillation is located a distance $2R$ below the pivot.
>
> The center of percussion is located at the same point.

8-29

A "hula hoop" of mass 200 grams and radius 50 cm is nailed to the wall at point A so that it can swing freely in a vertical plane as a physical pendulum.
a) Prove that the moment of inertia of the hoop about a axis through A and perpendicular to the plane of the hoop is 0.1 kgm^2.
b) What would be the frequency, in Hertz, of small oscillations of the hoop?

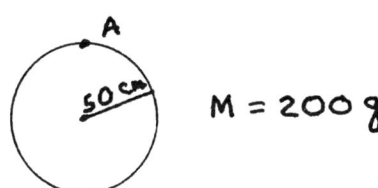

****

a) $I_{CM} = Mr^2 = 0.200(.500)^2 = 0.050$ kg·m^2

$I_A = I_{CM} + Mr^2 = 2 I_{CM} = 0.100$ kg·m^2

b) This is a physical pendulum

Let L_A be angular momentum about A

Let Γ_A be torque about A

$\Rightarrow \quad \dfrac{dL_A}{dt} = \Gamma_A$

$L_A = I_A \dfrac{d\theta}{dt} \qquad \Gamma_A = -mgr \sin\theta$

$I_A \dfrac{d^2\theta}{dt^2} = -mgr \sin\theta$

Assume: $\sin\theta \approx \theta$

$\dfrac{d^2\theta}{dt^2} = -\dfrac{Mgr}{I_A}\theta$

Solutions are:
$$\theta = \theta_0 \sin(\omega t + \phi)$$

with $\omega = \sqrt{\dfrac{Mgr}{I_A}} = \sqrt{\dfrac{(0.2)(9.8)(0.5)}{0.1}} = 3.13 \dfrac{rad}{s}$

$f = \dfrac{\omega}{2\pi} = 0.498$ Hz

8-30

A physical pendulum made from a meter stick, pivoted 0.25 m from one end, oscillating with small amplitude, and a spring oscillator of mass m = 1.0 kg, are to have the same period. Compute the necessary spring constant, and compute the period.

***

For the meter stick, $I = I_{CM} + mh^2$, where $I_{CM} = \dfrac{1}{12} m l^2$; $l = 1$ m, and $h = 0.25$ m; and $\omega^2 = mgh/I$,

$\omega^2 = \dfrac{gh}{\frac{1}{12}l^2 + h^2} = \dfrac{9.8 \times 0.25}{\frac{1}{12} + 0.25^2} = 16.8/sec^2$.

For the spring, $k = m\omega^2 = \underline{16.8\ kg/sec^2}$.

Period: $T = 2\pi/\omega = 2\pi/\sqrt{16.8} = \underline{1.53\ sec}$

9
FLUIDS

PRESSURE AND DENSITY IN FLUIDS

━━━ 9-1

An atmospheric pressure of 10^5 N-m^{-2} corresponds to a 753 mm mercury column. Suppose a certain person has a heartbeat of 80 per minute and that each beat displaces 50 cm^3 of blood at an average pressure difference of 120 mm Hg. Find the average power output of this person's heart.

****

$$\frac{10^5}{\Delta P} = \frac{753}{120}, \quad \Delta P = \frac{120}{753} \times 10^5 \frac{N}{m^2} = 1.59 \times 10^4 \frac{N}{m^2}$$

$$P = \vec{F} \cdot \vec{v} = PA \frac{d\ell}{dt} = P \frac{dV}{dt} \quad \text{where } dV = A \, d\ell$$
$$V \equiv \text{Volume}$$

$$P = 1.59 \times 10^4 \times \frac{50 \times 10^{-6} \, m^3}{0.75 \, s} \quad \Delta t = \frac{1}{80} \frac{\min}{\text{beat}} \times 60 \frac{s}{\min}$$

$$P = 1.06 \text{ Watt}$$

9-2

A mercury manometer is connected to a flask as shown in the diagram below. The the height difference in the manometer is 18 cm, what is the pressure within the flask? The mass density of mercury is 13.6 g/cm³.

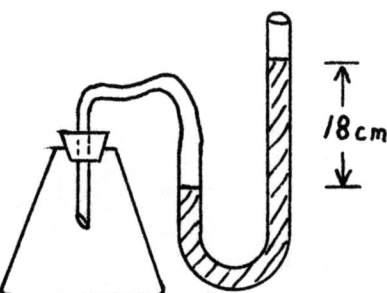

***

Let p_0 = 1 atm = pressure on open end of manometer

$p_B = p_0 = 1\,atm = 1.01 \times 10^5\,Pa$ (pascals)

Since A and C are at the same level

$p_A = p_C$ p_A = pressure within flask

$p_C = p_B$ + pressure due to weight of fluid

$p_C = p_B + \rho g h$ where $\rho = 13.6 \frac{g}{cm^3} = 13600 \frac{kg}{m^3}$

$\Rightarrow p_A = p_0 + \rho g h = 1.01 \times 10^5 + (13600)(9.8)(.18)$

$p_A = 1.25 \times 10^5\,Pa$

9-3

Pressure And Density In Fluids

A U-shaped hollow tube of cross sectional area, A, is filled with a liquid with density ρ. The fluid is momentarily disturbed so that the level in the right arm rises by an amount y, and hence lowers by an amount y in the left arm.

(a) Show that the liquid will undergo simple harmonic motion. (Ignore surface tension, viscosity and other damping effects.)

(b) Derive an expression for the frequency of the oscillation.

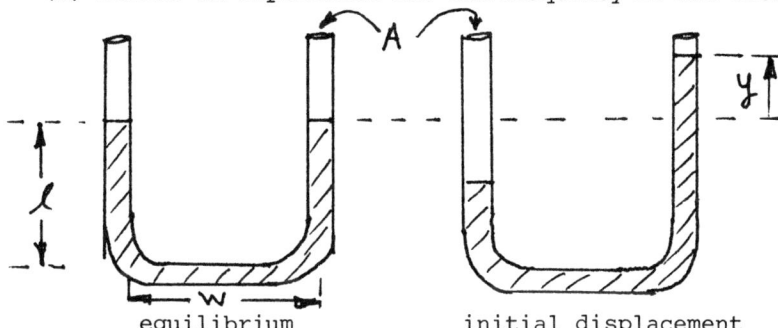

equilibrium   initial displacement

****

(a) & (b) together:

The pressure difference between the left and right hand arms when the fluid is displaced is
$$\Delta P = \rho g (2y)$$

Thus the unbalanced force on the displaced fluid is
$$F_{net} = \Delta P A = -(2yA)\rho g$$

The total mass of the fluid in motion is
$$Mass = \rho \, Volume = \rho \, (length \times cross\ section)$$
$$= \rho (2\ell + w) A$$

From Newton's 2nd Law: $F_{net} = m\ddot{y}$ or
$$-(2yA)\rho g = \rho (2\ell + w) A \ddot{y}$$

giving
$$\ddot{y} + \left(\frac{2g}{2\ell + w}\right) y = 0$$

which is the equation for simple harmonic motion with angular frequency $\omega = \sqrt{\dfrac{2g}{2\ell + w}}$

ARCHIMEDES' PRINCIPLE

9-4

A ping-pong ball of radius 2 cm and mass 5 g is released near the bottom of a tank of water. What is the initial upward acceleration of the ball? The mass density of water is 1 g/cm^3.

* * *

force diagram

$a \uparrow \quad \uparrow F_B$
$\quad \quad \bullet$
$\quad \quad \downarrow mg$

F_B = buoyant force

$\Sigma F = F_B - mg = ma$

Archimedes' Principle:

F_B = weight density of water × submerged volume of object

$F_B = \rho g \times \frac{4}{3} \pi R^3 \quad\quad \rho = 1 \frac{g}{cm^3} = 1000 \frac{kg}{m^3}$

$\quad\; = (1000)\, 9.8 \times \frac{4}{3} \pi (.02)^3 \quad R = .02\, m$

$F_B = .328\, N$

$\Rightarrow \quad F_B - mg = ma$

$\quad\quad a = \frac{1}{.005}\left(.328 - .005(9.8)\right) = 55.9\, \frac{m}{s^2}$

9-5

A ship weighing one million pounds contains 500,000 pounds of scrap iron in its hold. (Density of water = 62.4 pounds/ft^3 ; specific gravity of iron = 7.8). The ship floats in a canal lock. (a) What volume of water is displaced? (b) The scrap iron is now thrown overboard. Find the total volume of water now displaced by the ship and the iron. According to your results, what (if anything) happens to the water level in the canal lock?

****

(a) Weight of ship plus iron = 1.5×10^6 pounds
  = Buoyant force = Weight of water displaced

$$V = \frac{1.5 \times 10^6 \text{ pounds}}{62.4 \text{ pounds/ft}^3} = 2.4 \times 10^4 \text{ ft}^3 \text{ of water displaced}$$

(b) V (displaced by ship) = $\frac{1.0 \times 10^6 \text{ pounds}}{62.4 \text{ pounds/ft}^3}$ = 16000 ft^3

Density of iron = (specific gravity) × (Density of water)
  = 7.8 × 62.4 = 487 pounds/ft^3

V (displaced by iron) = V_{iron}

$$\frac{5 \times 10^5 \text{ pounds}}{487 \text{ pounds/ft}^3} = 1030 \text{ ft}^3 \text{ displaced by iron}$$

Total volume displaced = 16000 + 1030
  = 17030 ft^3

Since 24000 ft^3 were displaced previously, the water level will drop when the iron is thrown overboard.

9-6

A solid iron weight is to be held in equilibrium under water by a wooden float. Both objects are completely immersed. Find the ratio of the volume of wood to the volume of iron.

Specific Gravity	
Water	= 1.0
Iron	= 7.8
Wood	= 0.7

***

$$B_I + B_W = W_I + W_W, \quad \rho_L V_I g + \rho_L V_W g = \rho_I V_I g + \rho_W V_W g$$

$$V_W(\rho_L - \rho_W) = V_I(\rho_I - \rho_L), \quad \frac{V_W}{V_I} = \frac{\rho_I - \rho_L}{\rho_L - \rho_W}$$

$$\frac{V_W}{V_I} = \frac{7.8 - 1.0}{1.0 - 0.7} = 22.7$$

9-7

A 100 kg block of aluminum is suspended from a thin steel wire of cross-sectional area 3E-6 square meters. The density of aluminum is 2700 kg per cubic meter, and Young's modulus for steel is 2E11 N per square meter. Compute the strain in the wire, assuming that the aluminum block is under water.

***

By Archimedes' principle, the block is buoyed up by force $B = \rho_{H_2O} g V$, and $V = \frac{m}{\rho_{Al}}$. The load on the wire is thus $F = mg - B = mg\left(1 - \frac{\rho_{H_2O}}{\rho_{Al}}\right) = 100 \times 9.8 \times \left(1 - \frac{1000}{2700}\right) = 617$ N.

Wire strain: $\frac{\Delta \ell}{\ell} = \frac{F/A}{Y} = \frac{617}{3 \times 10^{-6} \times 2 \times 10^{11}} \doteq \underline{0.001}$

9-8

A rectangular barge measures 12.0 m wide X 30.0 m long X 7.00 m deep. It weighs 3,800,000 N when empty. How much load can it carry so that 1.00 m of the barge stays above the water (fresh water) line?

****

Free-body diagram:

(B = buoyant force.
L = load.
ρ = water density.
V = barge volume (underwater only))

$$\rho V g - L - mg = 0 \implies \Sigma F_y = 0$$

$$L = \rho V g - mg$$

$$L = \left(\frac{1000 \text{ kg}}{m^3}\right)(12.0 \text{ m})(30.0 \text{ m})(6.00 \text{ m})\left(\frac{9.81 \text{ m}}{\sec^2}\right) - 3.8 \times 10^6 \text{ N}$$

$$L = 1.74 \times 10^7 \text{ N}$$

(or mass of load = 1.77×10^6 kg)

9-9

A sphere of volume 3.0 ft^3 and weight density 48 lbs/ft^3 is tied to a massive anchor by a 10-ft rope and dropped into sea water of weight density 64 lb/ft^3. The anchor falls 50 ft to the sea bottom and the sphere floats 10 ft above it, held there by the rope.

(a) What is the water pressure (not considering atmospheric pressure at the surface) at the level of the sphere?
(b) What is the tension in the rope?

* *

"Weight density", w, means Weight/Volume = $g \times$(Mass)/Volume
= $g\rho$, where ρ = Mass/Volume = (mass) density

(a) The pressure at a vertical distance, h, beneath the horizontal free surface of a liquid of density ρ is $\rho g h$ plus the air pressure at the surface (here to be disregarded).

h for sphere is $50 - 10 = 40$ ft.

pressure at sphere = $hg\rho = hw = 40$ ft $\times 64$ lb/ft^3
$= 2560$ lb/ft^2

(b) There are three forces on the sphere:
(i) the tension, T, of the rope; to be found.
(ii) Earth's gravitational force, its weight, W.
(iii) the resultant force of the water, B

W = (weight density) $\times$ Volume
= 48 lb/ft$^3 \times 3.0$ ft$^3 = 144$ lb, down

By Archimedes' Principle, the resultant force of the water is an upward buoyant force, B, equal to the weight of the 3.0 ft^3 of water which the sphere displaces:

B = (weight density of water) $\times$ (Volume)
= 64 lb/ft$^3 \times 3.0$ ft$^3 = 192$ lb, up

Since the sphere is in equilibrium:
Total upward force = Total downward force
$B = T + W$, or $192 = T + 144$
$\therefore T = 192 - 144 = 48$ lb.

9-10

What is the area of the smallest block of ice 1 m thick that will support a car of 2000 kg? The specific gravity of ice is 0.92 and it is floating in fresh water.

****

Describing problem:

$h \{$ ——— ice ——— water

$m(car) = 2000 \text{ kg}$
$\rho_w = 1000 \text{ kg/m}^3$
$\rho_I = 920 \text{ kg/m}^3$
$h = 1 \text{ m}$

Setting up eqs.

$$B = W_I + W_c \quad \text{— buoyant force}$$

$$\Rightarrow \rho_w V = m_I + m_c$$

$$\Rightarrow \rho_w V = \rho_I V + m_c$$

$$\Rightarrow (\rho_w - \rho_I) V = m_c$$

$$\Rightarrow (\rho_w - \rho_I) h A = m_c$$

$$\Rightarrow A = \frac{m_c}{(\rho_w - \rho_I) h}$$

Substituting numbers:

$$A = \frac{2000}{(1000-920)(1)} = \underline{\underline{25 \text{ m}^2}}$$

9-11

An iron cube floats in a bowl of liquid mercury at $0^\circ C$. (a) If the temperature is raised $30^\circ C$, will the cube float higher or lower in the mercury? (b) By what percent will the fraction of volume submerged change? Show all work. Use the fact that the decrease in mass density is related to the initial mass density, ρ_o, and increased temperature, Δt, by the formula:

$$\Delta \rho = -\rho_o \beta (\Delta t) \; ; \; \beta = \text{Coefficient of volume expansion.}$$

MATERIAL	MASS DENSITY, ρ_o	COEFFICIENT OF VOLUME EXPANSION, β
Mercury, Hg	1.36×10^4 kg/m³	1.80×10^{-4} 1/°C
Iron, Fe	7.80×10^3 kg/m³	3.50×10^{-5} 1/°C

****

(a) $\Delta \rho_{Hg} = -\rho_{o_{Hg}} \beta_{Hg} \Delta t = -(1.36 \times 10^4 \text{kg/m}^3)(1.8 \times 10^{-4} \text{1/°C})(30°C)$
$= -79.44 \text{ kg/m}^3$

The decreased mass density will allow the iron cube to float lower since the iron's density will not decrease, relatively speaking, as much as the mercury's density.

$\Delta \rho_{iron} = -\rho_{o_{Fe}} \beta_{Fe} \Delta t = -(7.8 \times 10^3 \text{kg/m}^3)(3.5 \times 10^{-5} \text{1/°C})(30°C)$
$= -8.190 \text{ kg/m}^3$

(b) Let us utilize Archimedes' Principle for this portion of the problem.

Before the temperature change: $\rho_{Fe} V_{Fe} g = \rho_{Hg} V_{sub} g$

(1) $V_{sub}/V_{Fe} = \rho_{Fe}/\rho_{Hg}$; V_{sub} = Submerged volume

After the temperature change: $(\rho_{Fe} + \Delta \rho_{Fe}) V_{Fe} = (\rho_{Hg} + \Delta \rho_{Hg}) V_{sub}^*$

(2) $V_{sub}^*/V_{Fe} = (\rho_{Fe} + \Delta \rho_{Fe})/(\rho_{Hg} + \Delta \rho_{Hg})$; V_{sub}^* = Submerged volume after the temperature change

From expressions (1) and (2) we have

$$\frac{V_{sub}}{V_{sub}^*} = \left[\frac{\rho_{Fe}}{\rho_{Hg}}\right] \left[\frac{\rho_{Hg} + \Delta \rho_{Hg}}{\rho_{Fe} + \Delta \rho_{Fe}}\right]$$

$$= \left[\frac{7.8 \times 10^3 \text{kg/m}^3}{1.36 \times 10^4 \text{kg/m}^3}\right] \left[\frac{1.36 \times 10^4 \text{kg/m}^3 - 79.44 \text{kg/m}^3}{7.8 \times 10^3 \text{kg/m}^3 - 8.19 \text{kg/m}^3}\right]$$

$= .9956$ or 99.56% $100\% - 99.56\% = \underline{0.44\%}$

9-12

A block of wood floats in water with two-thirds of its volume submerged, and in oil with nine-tenths of its volume submerged. Find the density of the wood and of the oil.

****

free body diagram:

B = buoyant force

$\Rightarrow$ buoyant force $B = \rho_{wood} \cdot V_{wood} \cdot g$

but from Archimedes' Principle, the buoyant force equals the weight of the displaced fluid $= (2/3\, V_{wood}) \rho_{H_2O} \cdot g$

$$\Rightarrow \tfrac{2}{3} V_{wood} \cdot \rho_{H_2O} \cdot g = \rho_{wood} \cdot V_{wood} \cdot g$$

$$\Rightarrow \boxed{\rho_{wood} = \tfrac{2}{3}\rho_{H_2O} = 0.67\ g/cm^3}$$

B = buoyant force = weight of displaced oil

$$= \rho_{oil} \cdot \left(\tfrac{9}{10} V_{wood}\right) g = \rho_{wood} \cdot V_{wood} \cdot g$$

$$\Rightarrow \boxed{\rho_{oil} = \tfrac{10}{9}\rho_{wood} = 0.74\ g/cm^3}$$

9-13

The hydrometer is an instrument used to measure the specific gravity (or relative density) of a liquid. Consider a hydrometer to be made from a closed cylindrical glass tube: radius = 2 cm, length = 24 cm. mass = 150 g. An additional mass of 50 g is placed at the bottom (but inside) the cylinder to insure that it floats vertically.

a) If placed in water of density 10^3 kg/m^3, what height, h, will be below the surface?

b) If placed in acid of specific gravity 1.25, what height will be below the surface?

c) What is the minimum specific gravity that is measurable with this hydrometer?

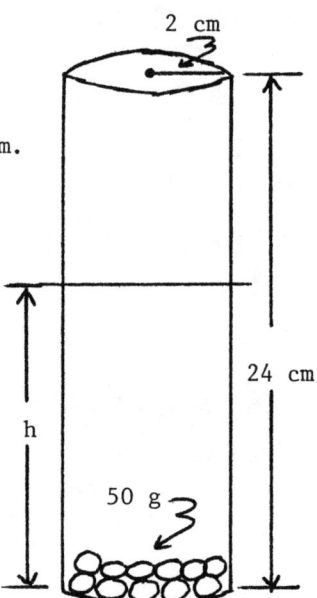

****

a) $\rho = \frac{m}{V} = 10^3 \text{ kg/m}^3 \quad m = 0.20 \text{ kg} \quad V = \pi r^2 h$

Wt displaced fluid = $\rho V g = mg$

$10^3 \frac{\text{kg}}{\text{m}^3} \times \pi \times (0.02\text{m})^2 \times h = 0.20 \text{ kg}$

$h = \frac{0.20 \text{ kg}}{\pi (0.02\text{m})^2 \times 10^3 \text{ kg/m}^3} = \underline{0.159 \text{ m}}$ (15.9 cm below surface)

b) $SG = 1.25 \quad \rho_{acid} = SG \times \rho_{water} = 1.25 \times 10^3 \text{ kg/m}^3$

$h = \frac{0.20 \text{ kg}}{\pi (0.02\text{m})^2 \times 1.25 \times 10^3 \text{ kg/m}^3} = \underline{0.127 \text{ m}} \quad (12.7 \text{ cm})$

c) Minimum specific gravity measurable occurs when entire cylinder is submerged, h = 24 cm.

$\rho = \frac{m}{V} = \frac{m}{\pi r^2 h} = \frac{0.20 \text{ kg}}{\pi (0.02\text{m})^2 \times 0.24 \text{ m}} = 663 \text{ kg/m}^3$

$SG = \rho/\rho_{water} = \underline{0.663}$

9-14

A solid block of wood of density 0.60 grams/cm³ floats on water of density 1.00 gram/cm³. Find the fraction of the volume of the block of wood which is above the surface of the water when the block is in static equilibrium.

****

Archimedes principle $\Rightarrow F_{BUOYANT} = m_{\substack{DISPLACED \\ FLUID}} \, g =$

$\rho_{FLUID} \, V_{\substack{DISPLACED \\ FLUID}} \, g = \rho_f V_f \, g.$

The weight of the block of wood $= m_{WOOD} \, g =$

$\rho_{WOOD} \, V_{WOOD} \, g = \rho_w V_w \, g.$

$\begin{array}{l} \uparrow F_{BUOYANT} = \rho_f V_f g \\ \\ \downarrow Weight = \rho_w V_w g \end{array} \Biggr\}$ TRANSLATIONAL Equilibrium $\Rightarrow$ NET FORCE ON WOOD BLOCK $= 0$.

So $\rho_f V_f g = \rho_w V_w g \Rightarrow V_f = \dfrac{\rho_w}{\rho_f} V_w = 0.6 \, V_w$

The fraction of the block above water

$= \dfrac{V_w - V_f}{V_w} = \dfrac{V_w - 0.6 V_w}{V_w} = \boxed{0.4 \text{ or } \tfrac{2}{5}}$

9-15

The figure shows two identical metal parallelepipeds, each suspended by a rope in a pool of water. Each has dimensions 30 cm x 30 cm x 60 cm. and a mass of 150 kilograms, but their orientations differ as shown. Find the tensions T_1 and T_2.

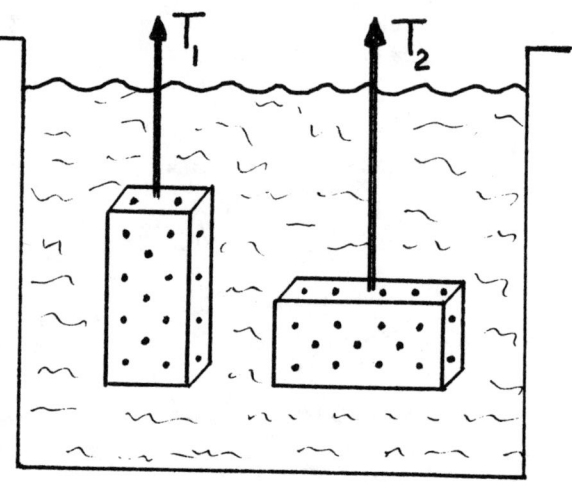

****

The tension and the buoyant force, B, act together to balance the block's weight, W:

So $T = W - B$.

a) Find $W = Mg = (150 \text{ kg})(9.8 \text{ m/s}^2) = \underline{1470 \text{ Nt}}$.

b) Find B: Archimedes $\Rightarrow$ B = weight of displaced water

$$= (\text{Water's density})(\text{volume})\, g$$

$$= (1000 \tfrac{kg}{m^3})(0.30 \times 0.30 \times 0.60) \times 9.8 \text{ m/s}^2$$

$\underline{B = 529 \text{ Nt}}$

So $T = W - B = 1470 \text{ Nt} - 529 \text{ Nt}$.

$\boxed{T = 941 \text{ Nt}}$

T_1 and T_2 are the same; all that counts is the mass and volume of each block, and those are the same.

9-16

A block of wood floats in water with 60% of its volume submerged. In a certain type of oil, the block floats with 90% of its volume submerged. Calculate the density of the wood and of the oil. The density of water is 1.0 gm/cc.

***

In both fluids, the buoyant force on the block is equal to the weight of the block. This buoyant force is equal to the weight of the volume of fluid displaced by the block.

Let ρ_B = block density, ρ_o = oil density, and ρ_W = water density.

$$F_B = m_w g = m_B g \quad \text{in water}$$

$$= (.60V)\rho_w g = V\rho_B g$$

$$\rho_B = .60 \rho_w = \underline{.60 \text{ gm/cc}}$$

$$F_B = m_o g = m_B g \quad \text{in oil}$$

$$= (.90V)\rho_o g = V\rho_B g$$

$$\rho_B = .90 \rho_o \quad \text{or} \quad \rho_o = \frac{\rho_B}{.90} = \frac{.60 \text{ gm/cc}}{.90}$$

$$\rho_o = \underline{.67 \text{ gm/cc}}$$

9-17

A block of wood of density .8 gm/cm³ and which has the shape of a cube 10 cm x 10 cm x 10 cm floats at the interface between a liquid and water as shown with its lower surface 4 cm below the interface. The density of water is 1 gm/cm³. (a) What is the density of the other liquid? (b) What is the pressure on the bottom surface of the block?

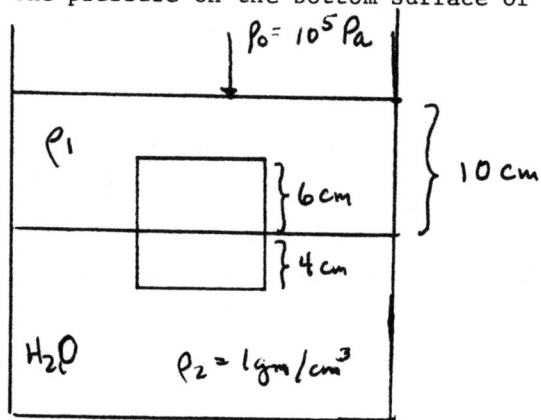

****

a) $BF = (.8 \text{ gm/cm}^3) g (1000 \text{ cm}^3)$

$= (1 \tfrac{gm}{cm^3}) g (4 \text{ cm})(100 \text{ cm}^2) + \rho_1 g (6 \text{ cm})(100 \text{ cm}^2)$

$800 \text{ gm} - 400 \text{ gm} = \rho_1 (600 \text{ cm}^3)$

$\rho_1 = \dfrac{400 \text{ gm}}{600 \text{ cm}^3} = .67 \text{ gm/cm}^3$

b) $P = P_0 + \rho_1 g h_1 + \rho_2 g h_2$

$= 10^5 \text{ Pa} + (.67 \tfrac{gm}{cm^3})(980 \tfrac{m}{s^2})(10 \text{ cm}) \dfrac{10^4 \text{cm}^2}{m^2} \times \dfrac{1 \text{ kg}}{1000 \text{ gm}}$

$+ (1 \tfrac{gm}{cm^3})(9.80 \tfrac{m}{s^2})(4 \text{ cm})(10^4 \tfrac{cm^2}{m^2}) \times \dfrac{1 \text{ kg}}{1000 \text{ gm}}$

$= 10^5 \text{ Pa} + 656.6 \text{ Pa} + 392 \text{ Pa}$

$P - P_0 = 1.05 \times 10^3 \text{ Pa}$

pressure is 1.05×10^3 Pa above atmospheric

9-18

How many people, each of mass 80 kg, can stand on an iceberg, of mass 4000 kg, before the iceberg is completely submerged? The density of water is 10^3 kg/m^3; the density of ice is $0.92 \times 10^3 \text{ kg/m}^3$.

****

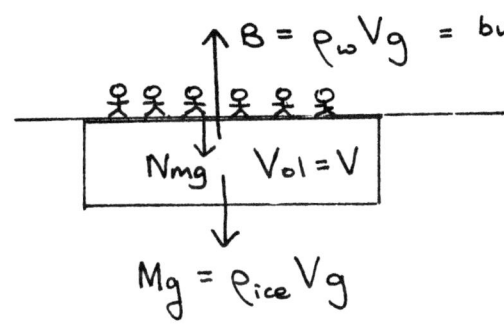

$B = \rho_w V g$ = buoyant force

$Mg = \rho_{ice} V g$

Let N be number of people possible, each of mass m.

for equilibrium: $\rho_{ice} V g + N m g = \rho_w V g$

$$\therefore N = \frac{(\rho_w - \rho_{ice}) V g}{mg}$$

but $V = \dfrac{M}{\rho_{ice}}$ $\therefore N = \dfrac{(\rho_w - \rho_{ice})}{m} \dfrac{M}{\rho_{ice}}$

$$\therefore N = \frac{\left(10^3 - 0.92 \times 10^3\right) \frac{kg}{m^3}}{0.92 \times 10^3 \frac{kg}{m^3}} \cdot \frac{4000 \text{ kg}}{80 \text{ kg}}$$

$\therefore N = 4.35$

$\therefore$ Only 4 people can stand on the iceberg.

9-19

When you step on the bathroom scales you do not measure your actual weight, but rather your actual weight minus the bouyant force of the atmosphere. If a person's apparent weight on the scales is 120.00 lb, what is this person's true weight?

Use the density of sea water, 64.27 lb/ft^3 as the approximate weight density of the person, and 1.3x10^{-3} as the specific gravity of air relative to fresh water (weight density 62.43 lb/ft^3).

****

Volume of person $V_{person} = \dfrac{mass}{\rho} = \dfrac{weight}{\rho g} = \dfrac{120 \text{ lb}}{64.27 \text{ lb/ft}^3}$

$V_{person} = 1.867 \text{ ft}^3$

Bouyant Force = weight of air displaced

$= \rho_{air} \, g \, V_{person} = (\text{sp. grav. air})(\rho_{water} \, g)(Vol)$

$= (1.3 \times 10^{-3})(62.43 \, \dfrac{lb}{ft^3})(1.867 \, ft^3)$

$= 0.15 \text{ lb}$

Apparent weight = Actual weight − Bouyant force

Actual weight = 120.00 lb + 0.15 lb = <u>120.15 lb</u>

PASCAL'S PRINCIPLE

9-20

A force acting on a piston at the top of a container causes an increase in pressure of the inclosed fluid which is at rest. Find the force exerted on the bottom of the container.

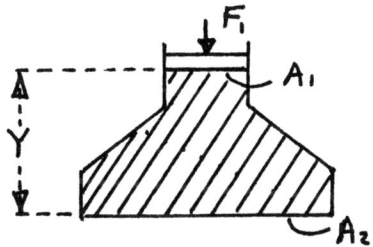

$F_1 = 100$ N
$A_1 = 10^{-4}$ m^2
$A_2 = 10^{-2}$ m^2
$\rho = 10^3$ kg/m^3
$Y = 1$ m
$g = 9.8$ m/s^2

* *

$$P_2 = P_1 + \rho g h \qquad P_2 = \frac{100}{10^{-4}} + 10^3(9.8)(1)$$

$$P_2 = 1.01 \times 10^6 \text{ N/m}^2 \qquad F_2 = (1.01 \times 10^6)(10^{-2}) = 1.01 \times 10^4 \text{ N}$$

FORCES AGAINST A DAM

9-21

A complex liquid fills a rectangular glass tank and is allowed to settle for many days. After this time the density of the liquid is nonuniform, increasing with depth according to the formula

$$\rho = \rho_o (1+y/d)$$

where y is the depth below the surface and d is the total depth of the liquid. Calculate
  (a) the pressure in the liquid as a function of y,
  (b) the total force on the end of the tank of width w, and
  (c) the torque on the end of the tank as measured from the top of the liquid.

****

(a) The weight of liquid in a horizontal slice of thickness Δy is:

$$\Delta W = (\rho g)(\ell w \Delta y)$$

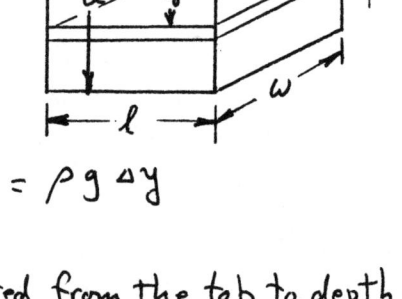

The increase in pressure due to this slice is:

$$\Delta P = \frac{\Delta W}{Area} = \frac{\rho g \ell w \Delta y}{\ell w} = \rho g \Delta y$$

So the total pressure accumulated from the top to depth y is:

$$P = \int_0^y \rho g\, dy' = g\int_0^y \rho\, dy' = g\rho_o \int_0^y (1+y'/d)\, dy'$$

$$= g\rho_o \left(y' + \frac{y'^2}{2d}\right)\Big|_0^y = \underline{g\rho_o \left[y + \frac{y^2}{2d}\right]}$$

(We ignore P_o the air pressure on the liquid surface for, although it acts on the inside wall, an equal pressure acts on the outside of the tank.)

(b) The total force against the end of width ω is:
$$F = \int P\,dA = \rho_0 g \omega \int_0^d \left(y + \frac{y^2}{2d}\right) dy$$
$$= \rho_0 g \omega \left(\frac{y^2}{2} + \frac{y^3}{6d}\right)\bigg|_0^d = \rho_0 g \omega \left(\frac{d^2}{2} + \frac{d^3}{6d}\right) = \underline{\rho_0 g \omega \frac{2d}{3}}$$

(c) The torque about the water surface is:
$$\tau = \int_0^d y\,dF = \int_0^d y(P\omega\,dy) = \rho_0 g \omega \int_0^d \left(y^2 + \frac{y^3}{2d}\right) dy$$
$$= \rho_0 g \omega \left(\frac{y^3}{3} + \frac{y^4}{8d}\right)\bigg|_0^d = \rho_0 g \omega \left(\frac{d^3}{3} + \frac{d^3}{8}\right)$$
$$= \underline{\rho_0 g \omega \left(\frac{11}{24} d\right)}$$

THE EQUATION OF CONTINUITY

9-22

ALL OF THE WATER FLOWING IN A LARGE STREAM IS DIRECTED TO FLOW THROUGH A PIPE, THE AREA OF WHICH DECREASES IN THE DOWNSTREAM DIRECTION, AND WHOSE CENTERLINE DOES NOT CHANGE ALTITUDE, AS SHOWN. THE EXITING FLOW OF WATER IS TO BE USED TO POWER A WATER TURBINE, WHICH REQUIRES THE WATER TO HAVE A VELOCITY OF 50 Ft/SEC. THE VELOCITY OF THE STREAM RANGES BETWEEN 5.0 AND 25 Ft/SEC. CROSS-SECTIONAL AREA OF THE STEAM IS 100 SQ. FT.

A) FIND THE RANGE OF AREA VALUES TO WHICH THE EXIT AREA MUST BE ADJUSTABLE TO ALLOW PROPER OPERATION OF THE TURBINE.

B) FIND THE RANGE OF VALUES FOR THE EXIT AREA IF THE PIPE IS BENT DOWNWARD SO THAT ITS CENTERLINE ALTITUDE DECREASES BY 50 FEET.

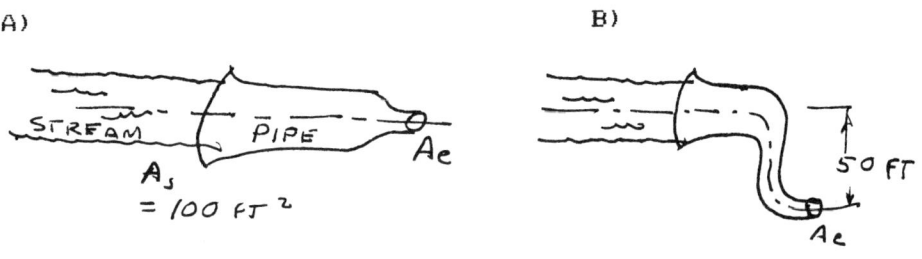

****

A) The flow of water through a pipe is governed by the continuity law, (or conservation of mass).

$$\rho_s A_s V_s = \rho_e A_e V_e$$

For liquids, density is essentially constant, therefore

$$A_s V_s = A_e V_e$$

When stream velocity = 5 ft/sec, the exit area must equal

$$A_e = A_s \frac{V_s}{V_e} = 100 \, FT^2 \times \frac{5 \, FT/sec}{50 \, FT/sec} = 10 \, FT^2$$

When stream velocity = 25 ft/sec, the exit area must equal

$$A_e = 100 \, FT^2 \times \frac{25 \, FT/sec}{50 \, FT/sec} = 50 \, FT^2$$

B) Since the continuity equation is not a function of altitude, this factor causes no change to the results.

BERNOULLI'S EQUATION

9-23

A house has a flat roof of area $400 m^2$. The windows of the house are tightly closed such that the air pressure in the house remains at the level $10^5 N\text{-}m^{-2}$ with density $1.3 kg\text{-}m^{-3}$, more-or-less independent of what happens outside. A tropical storm results in the sudden onset of a horizontal wind over the roof at a ground speed of $20 m\text{-}s^{-1}$. The roof, which is not nailed to the walls, is slightly lifted during the high wind.

a) Name the physical principle that explains the lifting of the roof and give the approximate time of its discovery.
b) Find the approximate mass of the roof.
c) After the storm, the roof, which contains one-fourth of the mass of the house, is repositioned and nailed to the walls. How fast must a new storm blow to lift the entire house?

****

a.) BERNOULLI'S PRINCIPLE — Early 18th century

b.) $P_{inside} = P_{outside} + \frac{1}{2}\rho v_{outside}^2$

$P = P_{in} - P_{out} = \frac{1}{2}\rho v_0^2 = \frac{1}{2} \times 1.3 \times (20)^2 = 260 \frac{N}{m^2}$

$m = \frac{W}{g} = \frac{PA}{g} = \frac{260 \times 400}{9.8} = 1.06 \times 10^4 kg$

c.) $m_{house} = 4 \, m_{roof} \quad \therefore F_{house} = 4 F_{roof}$

$\therefore \frac{1}{2}\rho v_{house}^2 = \frac{4}{2}\rho v_{roof}^2$

$v_{house} = 2 v_{roof} = 2 \times 20 = 40 \, m/s$

9-24

A pipe has a cross-sectional area of 2 cm² and then expands to 4 cm² at an elevation of 5 m above the original height as shown below. An incompressible fluid of density 600 kg/m³ flows through the pipe and has a speed of 20 m/s where the area is 2 cm². If the pressure = 9 x 10⁵ Pa where the area is 2 cm², what is the pressure where the area is 4 cm²?

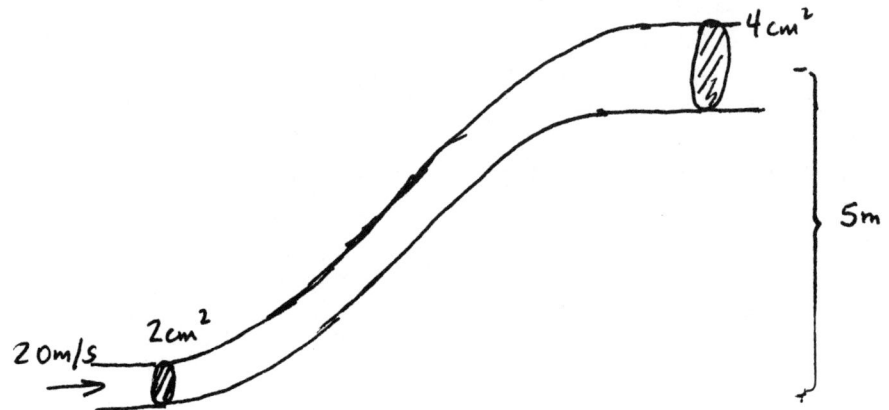

****

$$\rho_1 v_1 A_1 = \rho_2 v_2 A_2$$

$$v_2 = \frac{(20 m/s)(2 cm^2)}{4 cm^2} = 10 m/s$$

$$P_1 + \rho g y_1 + \tfrac{1}{2}\rho v_1^2 = P_2 + \rho g y_2 + \tfrac{1}{2}\rho v_2^2$$

$$P_2 = 9 \times 10^5 \, Pa + (600 \tfrac{kg}{m^3})(9.8 \tfrac{m}{s^2})(-5m)$$
$$+ \tfrac{1}{2}(600 \tfrac{kg}{m^3})[(20 m/s)^2 - (10 m/s)^2]$$

$$= 9 \times 10^5 \, Pa - .294 \times 10^5 \, Pa + .9 \times 10^5 \, Pa$$

$$P_2 = 9.61 \times 10^5 \, Pa$$

9-25

A car moves with constant speed along a horizontal circular path.
(a) Find the equilibrium angle between a simple pendulum and the vertical.
(b) Find the period of small amplitude oscillations.

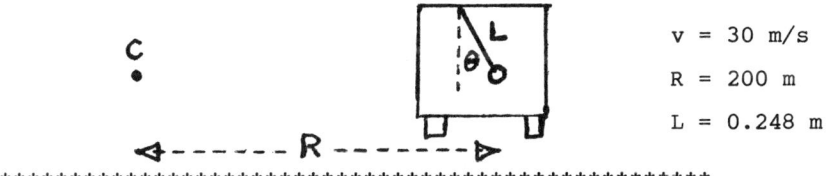

$v = 30$ m/s
$R = 200$ m
$L = 0.248$ m

(a) Inside the car, the effective field, $\vec{g}'$, is the vector sum of $\vec{g}$ and $-\vec{a}_r$, where $\vec{a}_r$ is the centripetal acceleration. $v^2/R = \dfrac{30^2}{200} = 4.5$ m/s^2.

$$\tan\theta = \frac{a_r}{g} = \frac{4.5}{9.8}$$

$$\boxed{\theta = 24.7°}$$

CHECK: From the viewpoint of an observer outside the car, in an inertial reference frame, the only forces acting on the pendulum bob are the string tension F, and the force of gravity mg. F_c is the resultant force toward the center of the circle, C.

$$\tan\theta = \frac{F_c}{mg} = \frac{mv^2/R}{mg}$$

$$\tan\theta = a_r/g.$$

(b) $g' = \sqrt{g^2 + a_r^2} = \sqrt{(9.8)^2 + (4.5)^2} = 10.78$ m/s^2

$$T = 2\pi\sqrt{\frac{L}{g'}} = 2\pi\sqrt{\frac{0.248}{10.78}} = \boxed{0.953 \text{ s}}$$

9-26

A venturi tube is mounted vertically as shown. The diameter D of the larger tube section is three times the diameter of the smaller tube section (i.e. D=3d). The pressure difference in the two sections of the tube is measured by a U tube. The difference in elevations where the U tube is connected to the venturi tube is $\Delta H = 4 \cdot \Delta h$ where Δh is the difference in elevation of mercury in the U tube. If $\Delta h = 2$cm find the velocity v_1 of the water in the upper section of the venturi tube. The specific gravity of mercury is 13.5.

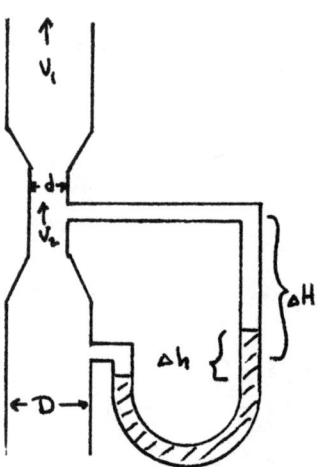

****

Setting up eqs.

$$P_1 + \tfrac{1}{2}\rho v_1^2 + \rho g h_1 = P_2 + \tfrac{1}{2}\rho v_2^2 + \rho g h_2$$

$$\Rightarrow \tfrac{1}{2}\rho(v_2^2 - v_1^2) = P_1 - P_2 + \rho g(h_1 - h_2)$$

Using the continuity eq. $A_1 v_1 = A_2 v_2$

$$\tfrac{1}{2}\rho\left(v_1^2 \tfrac{A_1^2}{A_2^2} - v_1^2\right) = \Delta P - \rho g \Delta H$$

$$\Rightarrow v_1^2 = \frac{2(\Delta P - \rho g \Delta H)}{\rho\left(\tfrac{A_1^2}{A_2^2} - 1\right)}$$

But from U-tube $\Delta P = \rho_{Hg} g \Delta h - \rho g \Delta h$
$$= g \Delta h (\rho_{Hg} - \rho)$$

$$\Rightarrow v_1^2 = \frac{2[g\Delta h(\rho_{Hg} - \rho) - \rho g \Delta H]}{\rho\left(\tfrac{A_1^2}{A_2^2} - 1\right)}$$

$$= \frac{2g[\Delta h(\rho_{Hg} - \rho) - \rho \Delta H]}{\rho\left(\tfrac{D^4}{d^4} - 1\right)}$$

$$= \frac{2g[\Delta h \rho_{Hg} - \Delta h \rho - 4\rho \Delta h]}{\rho\left[\tfrac{(3d)^4}{d^4} - 1\right]} = \frac{2g\Delta h(\rho_{Hg} - 5\rho)}{80\rho}$$

$$= \frac{2g\Delta h(\rho_{Hg}/\rho - 5)}{80}$$

So
$$V_1 = \sqrt{\frac{2g\Delta h(\rho_{Hg}/\rho - 5)}{80}}$$

Substituting numbers:

$$V_1 = \sqrt{\frac{2(9.8)(.02)(13.5-5)}{80}} = .2 \text{ m/s} = 20 \text{ cm/s}$$

9-27

Compute the lift suggested by Bernoulli's principle on a wing of area 50 m² if the air passes over the top and bottom surfaces at speeds of 320 m/sec and 290 m/sec respectively. Air density is 1.29 kg/m³.

****

$$P_1 + \tfrac{1}{2}\rho v_1^2 + \rho g h_1 = P_2 + \tfrac{1}{2}\rho v_2^2 + \rho g h_2$$

Assume that $\rho g(h_2-h_1) \ll P_1 - P_2$ Then,

$$P_1 - P_2 \cong \tfrac{1}{2}\rho(v_2^2 - v_1^2)$$ and the *lift force*

$$F = (P_1-P_2)A \cong \tfrac{1}{2}\rho A(v_2^2 - v_1^2)$$

$$F = \tfrac{1}{2}\left(1.29 \tfrac{kg}{m^3}\right)(50\, m^2)\left[\left(320 \tfrac{m}{sec}\right)^2 - \left(290 \tfrac{m}{sec}\right)^2\right]$$

$$\boxed{F = 5.9 \times 10^5 \, N}$$

To check the assumption that $\rho g(h_2-h_1) \ll P_1-P_2$, consider

$$h_2 - h_1 \ll \frac{P_1-P_2}{\rho g} = \frac{F}{\rho g A} = \frac{5.9 \times 10^5 N}{\left(1.29 \tfrac{kg}{m^3}\right)\left(9.8 \tfrac{m}{sec^2}\right)(50\, m^2)}$$

$\Rightarrow h_2 - h_1 \ll 933$ meters which is easily satisfied for ordinary wings.

9-28

Water enters the pipe shown below from the left at a speed of 5.00 m/s. The pressure on the left side is 2.00 atmospheres, and the cross-sectional area of the pipe is 0.100 m². The pipe is 0.500 meters higher on the right, and has an area of 0.200 m². Calculate the speed of the water in the pipe on the right, and its pressure.

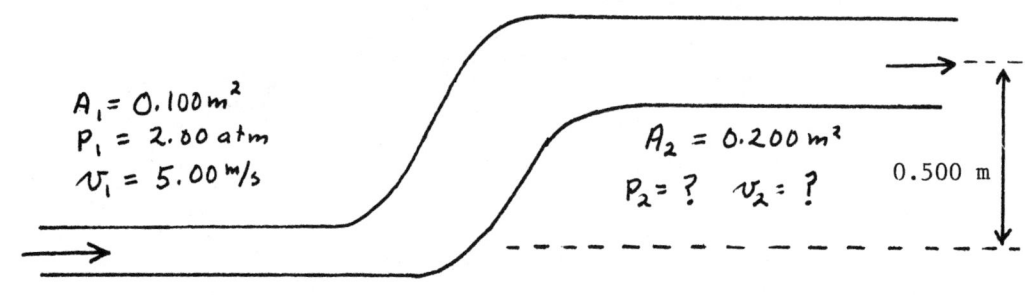

$A_1 = 0.100 \, m^2$
$P_1 = 2.00 \, atm$
$v_1 = 5.00 \, m/s$

$A_2 = 0.200 \, m^2$
$P_2 = ? \quad v_2 = ?$

0.500 m

****

Use continuity equation to get v_2:

$$\rho_1 v_1 A_1 = \rho_2 v_2 A_2$$

Since water is essentially incompressible $\rho_1 = \rho_2$.

$$v_1 A_1 = v_2 A_2$$

$$v_2 = \frac{v_1 A_1}{A_2}$$

$$v_2 = 2.5 \, m/s$$

Get P_2 from Bernoulli's Equation:

$$P_1 + \tfrac{1}{2}\rho v_1^2 + \rho g y_1 = P_2 + \tfrac{1}{2}\rho v_2^2 + \rho g y_2$$

$$P_2 = P_1 + \tfrac{1}{2}\rho(v_1^2 - v_2^2) + \rho g(y_1 - y_2)$$

$$= 2 \times 10^5 \, N/m^2 + 9375 \, N/m^2 - 4900 \, N/m^2$$

[Watch units! Remember 1 atm ~ $10^5 \, N/m^2$ & $\rho_{water} = 1000 \, kg/m^3$]

$$P_2 = 2.05 \times 10^5 \, N/m^2$$

Or about 2.05 atm

9-29

Water at a pressure of 4.3 atmospheres at street level flows into an office building at a speed of 0.50 m/sec through a pipe 5.50 cm in diameter. The pipes taper down to 2.5 cm in diameter by the top floor, 25 m above. Neglecting the viscosity, calculate the flow velocity and the pressure in such a pipe on the top floor. The pressure, 4.3 atmospheres, given above is the gauge pressure.

****

$A_1 = (0.055 m)^2 \pi/4 = 2.376 \times 10^{-3} m^2$
$A_2 = (0.025 m)^2 \pi/4 = 4.909 \times 10^{-4} m^2$

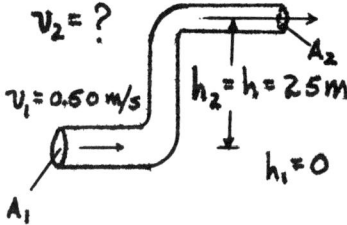

From the equation of continuity, the flow rate R is

$$R = v_1 A_1 = v_2 A_2$$

and

$$v_2 = \text{flow velocity} = v_1 A_1 / A_2 = \frac{(0.5 \, m/sec)(2.376 \times 10^{-3} m^2)}{4.909 \times 10^{-4} m^2}$$

$$= 2.420 \, m/sec$$

Let us now introduce Bernoulli's equation.

(1) $P_1 + \rho g h_1 + \frac{1}{2} \rho v_1^2 = P_2 + \rho g h_2 + \frac{1}{2} \rho v_2^2$

where $P_1 - P_{atm} =$ gauge pressure at the point where P_1 is measured, i.e. $P_{1\,gauge}$. Thus,

(2) $P_{1\,gauge} + \rho g h_1 + \frac{1}{2} \rho v_1^2 = P_{2\,gauge} + \rho g h_2 + \frac{1}{2} \rho v_2^2$

$$P_{1\,gauge} = 4.3 \, atm \left(\frac{1.013 \times 10^5 \, Pa}{1 \, atm} \right) = 435,590 \, Pa$$

From equation (2), $P_{2\,gauge}$ is expressed as

$$P_{2\,gauge} = P_{1\,gauge} - \rho g h + \frac{1}{2} \rho (v_1^2 - v_2^2)$$

$P_{2\,gauge} = 435,590 \, Pa - (1000 \, kg/m^3)(9.81 \, m/sec^2)(25 \, m)$
$\qquad\qquad + \frac{1}{2}(1000 \, kg/m^3)[(0.5 \, m/sec)^2 - (2.42 \, m/sec)^2]$

$= 187,536.8 \, Pa = 1.851 \, atm$

or
$P_2 = 288,800 \, Pa = 2.851 \, atm$

9-30

IN ORDER TO PROVIDE WATER UNDER PRESSURE IN A HIGH RISE APARTMENT BUILDING, A LARGE STORAGE TANK IS BUILT UPON THE ROOF. IT IS VENTED TO THE ATMOSPHERE. ASSUME NO ENERGY IS LOST BY THE WATER AS IT PASSES THROUGH THE PIPES.

A) FIND THE WATER SPEED AS IT LEAVES A FAUCET ON THE FIRST FLOOR AND ONE ON THE 50th floor.

B) WHEN THE FAUCETS ARE CLOSED, FIND THE WATER PRESSURE AT THE SAME 2 FAUCETS.

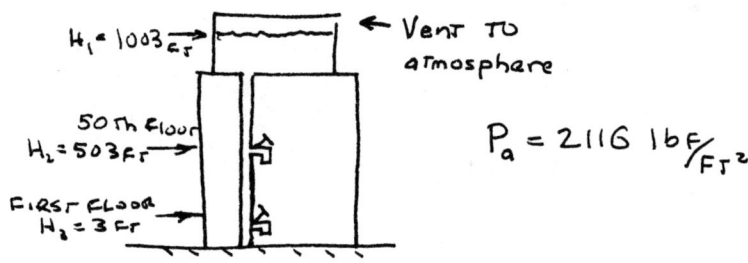

****

A) We have two equations for fluid flow problems, the continuity equation and Bernoulli's Equation.

1) $\rho_1 A_1 V_1 = \rho_2 A_2 V_2$

2) $P_1 + \frac{1}{2}\rho_1 V_1^2 + \rho_1 g h_1 = P_2 + \frac{1}{2}\rho_2 V_2^2 + \rho_2 g h_2$

The following values are given

P1 = ambient since it is the top of the water stack

P2 = ambient since pressure in a fluid exiting a pipe must equal ambient

Y1, Y2 and Y3, from the figure.

from 1) $V_1 = A_2/A_1 V_2 \approx 0$ since $A_2/A_1 << 1$

Therefore, solving for the velocity at the exit (from 2)

$$2116 \tfrac{lb_f}{ft^2} + \tfrac{62.4\, lb_m/ft^3}{2} \times 0 + 62.4 \tfrac{lb_f}{ft^3}(Y_1 - Y_F)$$
$$= 2116 \tfrac{lb_f}{ft^2} + \tfrac{62.4\, lb_m/ft^3}{2} V_F^2$$

whe Y_F & V_F are at faucer

$$\therefore V_F^2 = 2(Y_1 - Y_F)\, \tfrac{lb_f}{lb_m} \times \tfrac{32.2\, lb_m\, ft/sec^2}{lb_f}$$

Y-Yf = the difference in height between the top of the tank and the specific faucet.

Solving for each of the faucets

$$V_2 = \sqrt{(1003\,ft - 503\,ft) \times 2 \times 32.2\,ft/sec^2} = 179\,ft/sec$$

$$V_3 = \sqrt{(1003\,ft - 3\,ft) \times 2 \times 32.2\,ft/sec^2} = 253\,ft/sec$$

Note that these velocities are very large. This results from the POOR ASSUMPTION of no energy loss in the pipes.

B) Using Bernoulli's Equation, where the flow velocities are zero, the pressure at each faucet *is* found.

For faucet 2

$$2116\,lbf/ft^2 + 62.4\,lbf/ft^3 \,(1003\,ft - Y_F) = P_2$$

$$P = 33.3 E3\,lbf/ft^2 = 231\,psi$$

For faucet 3

$$P = 64.5 E3\,lbf/ft^2 = 448\,psi$$

9-31

An airplane has wings of area 40 m² each. When the airplane is flying horizontally air flows over the top surface of the wings at 250 m/s and over the bottom surface at 220 m/s. If the density of air is 1.3 kg/m³ calculate the mass of the airplane.

****

Bernouilli:

$$P + \rho g h + \tfrac{1}{2}\rho v^2 = const$$

$$\therefore P_{top} + \tfrac{1}{2}\rho v_{top}^2 = P_{bott} + \tfrac{1}{2}\rho v_{bott}^2$$

$$\therefore \Delta P = P_{bott} - P_{top} = \tfrac{1}{2}\rho (v_{top}^2 - v_{bott}^2)$$

$$Lift = 2 \cdot Area \cdot \Delta P = A\rho (v_{top}^2 - v_{bott}^2)$$
↑
2 wings

but   Lift = mg m = mass of plane

$$\therefore m = \frac{A\rho}{g}(v_{top}^2 - v_{bott}^2)$$

$$\therefore m = \frac{(40 m^2)(1.3 \frac{kg}{m^3})((250)^2-(220)^2) \frac{m^2}{s^2}}{9.8 \, m/s^2} = 74816 \, kg$$

9-32

Water is moving with a speed of 8 m/sec through a pipe with a cross-sectional area of 6 cm^2. The water gradually descends 5 m as the pipe decreases in area to 4 cm^2. The pressure at the upper level is 2.5x10^4 N/m^2. Find the speed of flow and the pressure at the lower level.

****

From the equation of continuity,
$$V_1 A_1 = V_2 A_2$$
$$(8 m/s)(6 cm^2) = V_2 (4 cm^2) \qquad V_2 = 12 \, m/sec$$

From Bernoulli's equation,
$$P_1 + \rho g h_1 + \tfrac{1}{2}\rho v_1^2 = P_2 + \rho g h_2 + \tfrac{1}{2}\rho v_2^2$$

[ρ = density of water = 1000 kg/m^3]

$$2.5 \times 10^4 + (1000)(9.8)(5) + \tfrac{1}{2}(1000)(8)^2$$
$$= P_2 + (1000)(9.8)(0) + \tfrac{1}{2}(1000)(12)^2$$
$$P_2 = 3.4 \times 10^4 \, N/m^2$$